T0231085

Internet Resources
on Weight Loss
and Obesity

THE HAWORTH INFORMATION PRESS®
Haworth Internet Medical Guides
M. Sandra Wood, MLS
Editor

The Guide to Complementary and Alternative Medicine on the Internet by Lillian R. Brazin

Internet Guide to Travel Health by Elizabeth Connor

Internet Guide to Food Safety and Security by Elizabeth Connor

Internet Guide to Cosmetic Surgery for Women by M. Sandra Wood

Internet Guide to Anti-Aging and Longevity by Elizabeth Connor

Internet Guide to Herbal Remedies by David J. Owen

Internet Guide to Medical Diets and Nutrition by Lillian R. Brazin

Internet Guide to Cosmetic Surgery for Men by M. Sandra Wood

Internet Resources on Weight Loss and Obesity by Lillian R. Brazin

Internet Resources on Weight Loss and Obesity

Lillian R. Brazin, MS, AHIP

CRC Press
Taylor & Francis Group
Boca Raton London New York

CRC Press is an imprint of the
Taylor & Francis Group, an **informa** business

For more information on this book or to order, visit
http://www.haworthpress.com/store/product.asp?sku=5398

or call 1-800-HAWORTH (800-429-6784) in the United States and Canada
or (607) 722-5857 outside the United States and Canada

or contact orders@HaworthPress.com

Published by

The Haworth Information Press®, an imprint of The Haworth Press, Inc., 10 Alice Street, Binghamton, NY 13904-1580.

PUBLISHER'S NOTE

The development, preparation, and publication of this work has been undertaken with great care. However, the Publisher, employees, editors, and agents of The Haworth Press are not responsible for any errors contained herein or for consequences that may ensue from use of materials or information contained in this work. The Haworth Press is committed to the dissemination of ideas and information according to the highest standards of intellectual freedom and the free exchange of ideas. Statements made and opinions expressed in this publication do not necessarily reflect the views of the Publisher, Directors, management, or staff of The Haworth Press, Inc., or an endorsement by them.

This book has been published solely for educational purposes and is not intended to substitute for the medical advice of a treating physician. Medicine is an ever-changing science. As new research and clinical experience broaden our knowledge, changes in treatment may be required. While many potential treatment options are made herein, some or all of the options may not be applicable to a particular individual. Therefore, the author, editor and publisher do not accept responsibility in the event of negative consequences incurred as a result of the information presented in this book. We do not claim that this information is necessarily accurate by the rigid scientific and regulatory standards applied for medical treatment. **No warranty, express or implied, is furnished with respect to the material contained in this book. The reader is urged to consult with his/her personal physician with respect to the treatment of any medical condition.**

Due to the ever-changing nature of the Internet, Web site names and addresses, though verified to the best of the publisher's ability, should not be accepted as accurate without independent verification.

Cover design by Marylouise E. Doyle.

Library of Congress Cataloging-in-Publication Data

Brazin, Lillian R.
    Internet resources on weight loss and obesity / Lillian R. Brazin.—1st ed.
        p. cm.
    Includes index.
    ISBN-13: 978-0-7890-2649-1 (hard : alk. paper)
    ISBN-10: 0-7890-2649-X (hard : alk. paper)
    ISBN-13: 978-0-7890-2650-7 (soft : alk. paper)
    ISBN-10: 0-7890-2650-3 (soft : alk. paper)
    1. Weight loss—Computer network resources. 2. Obesity—Computer network resources. I. Title.

RM222.2.B69 2007
613.2'50285—dc22
                                                                                                        2006034454

To Fudge, a formerly obese cat, who lost his "mud flaps"
by following a lower-carbohydrate diet—
and laying off the dog's kibble!

ABOUT THE AUTHOR

Lillian R. Brazin, MS, AHIP, is Director of Library Services at the Albert Einstein Healthcare Network in Philadelphia. She has over thirty-five years of experience as a medical reference librarian, working in the areas of online services, user instruction, and management of academic health sciences libraries. She has eleven years of experience as an evening and weekend reference librarian for the Free Library of Philadelphia. Brazin has published numerous articles and developed and taught workshops on Internet resources. She is a member of the Academy of Health Information Professionals (Distinguished Level), American Library Association, Special Libraries Association, and Medical Library Association.

CONTENTS

Preface

A few years ago, the hospital where I am employed engaged two physicians from a prominent medical school to teach a series of continuing medical education workshops for physicians. You can imagine my surprise when I joined the two physicians for lunch in the hospital cafeteria. Both physicians were more than mildly overweight. They bypassed the salad bar and the sandwich line and ordered the infamous Philly cheese steak sandwich, minus the roll. I assumed they wanted to sample the local Philadelphia cuisine, but they told me that were on a very-low-carbohydrate diet. They planned to "kick-start" their weight loss by following the popular diet for a few weeks, and then would switch to a more traditional, balanced diet program. Being a medical snob and a cynic, I was taken by surprise because I looked down on diet groupies, viewing them as fools who followed whatever diet plan was in vogue. I thought about the doctors' plan for a few hours and decided that, if "Doctors Big Name Medical School" considered the trendy diet is safe to follow for a few weeks, then I could try it, too. At the time, a surprisingly large group of my hospital's physicians, nurses, dietitians, and other staff were also following one or the other highly publicized low-carbohydrate diets.

After checking with my family doctor, who asked me to get a few baseline blood tests, I embarked on a regimen of blue cheese omelets, bacon, ricotta with vanilla extract, and tins of sardines. I ate salads, too. The weight came off easily, but my total cholesterol rose to 280! You know it is serious when your physician writes you a note on the lab test results sheet, marks the abnormal blood levels, and mails it to you (even though my scheduled appointment was in less than a month). Daily oatmeal, hot flax meal cereal, and a moratorium on whole eggs, blue cheese, Brie, sardines, shrimp, and bacon quickly brought down the cholesterol count, although

Internet Resources on Weight Loss and Obesity
© 2007 by The Haworth Press, Inc. All rights reserved.
doi:10.1300/5398_a

there is still room for improvement. Of course, I gained back the excess weight, too. What happened to me may not happen to you. Not everyone who eats whole milk and egg-rich foods sees his or her total cholesterol level rise. It is wise, however, to know what your prediet blood values and blood pressure levels are before beginning any type of diet. If your blood glucose level was hovering in the diabetic range before starting a diet program, but dropped to the normal range after changing your diet and exercise habits, would you not feel encouraged and maintain these new habits?

Carefully read each section in the book. Try to be realistic about your weight and the commitment you must make to eating sensibly and exercising more frequently. I hope you find a diet plan that meets your needs for a lifetime of health and high energy. Good luck!

Acknowledgments

With much appreciation to the editors and production assistants who coaxed me through the long gestation period that resulted in the "birth" of this book.

Internet Resources on Weight Loss and Obesity
© 2007 by The Haworth Press, Inc. All rights reserved.
doi:10.1300/5398_b

Chapter 1

Advice for the Weight Challenged

INTRODUCTION:
DEFINING THE PROBLEM

Are you long past the age of baby fat? Is your body less lush than zaftig, less buff than beefy? Does your physique feature a beer barrel gut instead of six-pack abs? Would high-school classmates recognize you after 30 years? Can you face your old beau or girlfriend? There are many reasons to want to lose weight, but the superficial reasons are not why I wrote this book. Serious overweight, termed *obesity*, is linked to major health problems: type 2 diabetes, heart disease, high blood pressure, blood cholesterol abnormalities (dyslipidemia), stroke, gastroesophageal reflux disease, stress urinary incontinence, sleep apnea syndrome, and gall bladder disease. Obesity has been associated with some types of cancer, aggravates arthritis of the joints and varicose veins, complicates surgery, and may contribute to infertility, fatigue, and depression. A study, reported in the *Journal of the American Medical Association* in 2006, found that midlife obesity significantly increases the risk of hospitalization and death from congestive heart disease and diabetes, regardless of other risk factors.[1]

One-third of adult Americans are overweight (body mass indexes [BMI] of 25-29), and another third are obese (BMIs of 30 and greater). According to the National Childhood Obesity Foundation, 15 percent of children and teens, aged six to nineteen, are overweight, as are 10 percent of children between ages two and five. Annually, there are over 300,000 obesity-related deaths in this country. In the United States, obesity-related illnesses lead to annual medical costs of $90 billion.[2] Many Americans are on

Internet Resources on Weight Loss and Obesity
© 2007 by The Haworth Press, Inc. All rights reserved.
doi:10.1300/5398_01

weight-loss diets. Surgical procedures for weight reduction (bariatric surgery) are performed in most hospitals, often at bariatric surgery centers of excellence staffed by physicians and nurses who specialize in these surgical procedures. In the United States, an estimated 140,000 gastric bypass procedures were performed in 2004. Prospective patients wonder if such surgery is a quick solution to the problem of overweight. How do you locate and evaluate an expert bariatric surgeon? Is it best to go to a hospital-based bariatric surgery center of excellence, to a general surgical department within a local hospital, or to a private clinic? What are the side effects and risks, both immediate and long term, of weight-loss surgery procedures? How successful is the surgery? What is the difference between obesity and overweight? How do you calculate your BMI, the ratio used to determine whether a person is obese or overweight?

You may have been advised by your healthcare provider to follow a particular food plan or eliminate certain foods from your diet to aid weight loss. How do you locate up-to-date information about specific weight-loss diets? How do you find varied and tasty recipes that follow the guidelines and restrictions of your diet? Do you want to spend a lot of money purchasing a shelf full of cookbooks and nutrition-advice books? There should be a small reference library in every home, but after a year or two, some of the material will be obsolete or your tastes and lifestyle may have changed. Web sites for major diet programs are kept current, reflecting new nutritional research and new features, such as recipes, Q&A with dietitians, and success stories.

Many dieters enjoy interacting with others following the same weight-loss plan. One of the key features of the Weight Watchers Program is the group meeting, but not everyone has the time to attend regular meetings. Some of us are uncomfortable attending meetings or having (somewhat) public "weigh-ins." The Weight Watchers Web site has a fee-based online program, with discounts given to traditional meeting attendees. Other diet programs feature free online forums.

WHY YOU NEED TO READ THIS BOOK

This book covers the best Web sites to help you understand and solve the problem of obesity. There are sections to aid inexperienced Internet users and readers lacking medical backgrounds to learn to find and evaluate medical and nutrition information in cyberspace. What are the guidelines to follow when sifting through information found on the Web? You will learn:

- How to determine whether medical and nutrition information is correct
- How to locate recommended Web sites
- Where to begin researching particular diets or weight-loss methods
- How to use specific criteria to evaluate a Web site
- How to detect medical fraud
- When and how to use search engines
- What is the significance of "domains" in Web site addresses
- How to observe proper etiquette while participating in Internet discussion groups

Commercial sites were excluded if they did not include, free of charge, basic information about the diet program or method. Some sites, although maintained by board-certified physicians, exist primarily to sell the physicians' books or draw clients to their medical practice. Physicians with expertise outside of fields related to nutrition or to the diseases supposedly remedied by the diet regimens recommend certain books or diet philosophies. On one site, a plastic surgeon and an orthopedic surgeon write glowingly about a diet guru's books and diet plans. At this same Web site, little information is disclosed about the ground rules of the diet. A reader would have to enroll (for a fee) in the diet program or purchase the author's books. Follow your instincts. If a diet regimen sounds bizarre to you, do not even think about following it. Some diets are dangerously restrictive or promoted by people with questionable credentials (the diet proponents almost "quack"). Every city has expert physicians and certified dietitians. There are medical reporters at most television and radio stations and newspapers. Heed their advice.

Remember that the Internet is dynamic: Web sites appear, disappear, or change address or arrangement. At the time this book was published, all Web site addresses were current. In most cases, if a Web site moves to a different location, there will be a "forwarding address," or the old site will automatically link readers to the new one. If this does not happen, try using a search engine to search for the name of the Web site.

Sometimes a site is rearranged, and topics will appear on different "pages" of the Web site. Using the site's index should point you to the section you seek.

Do not forget to bookmark your favorite sites. You may find additional Web resources on the topics in this book, so add those to your folder of favorites.

NOTES

1. Yan, L.L., Daviglus, M.L., Liu, K., et al. (2006). Midlife body mass index and hospitalization and mortality in older age. *Journal of the American Medical Association* 295:190-198.

2. Vastag, B. (2004). Obesity now on everyone's plate. *Journal of the American Medical Association* 291:1186-1188.

Chapter 2

Hunting for Information

SEARCH ENGINES

Search engines allow you to type a word, phrase, or name, and retrieve Web sites and discussion group responses containing these words, phrases, or names. Depending on the search engine, the results that come up may or may not be relevant to your needs. Seek quality, not quantity. In most instances, you should not need to go on cyber fishing expeditions to locate useful Web sites. Use search engines to zero in on the correct URL (Web site address) of a recommended site or a resource mentioned in the newspaper or on television. For this reason, there will be no instruction on how to construct elaborate search strategies using search engines. Just keep it simple. In most instances, type a word or phrase in the white "search" block and click the Enter or Go button. Some of the search engines have topical directories that help narrow the search. Begin with the preselected groups of Web sites in the directories.

About.com
<http://www.about.com>

About.com began in 1997 as The Mining Company. In 2005, the New York Times Company acquired the site. About.com features two dozen general guides written by approximately 500 outside experts who have been selected and trained to manage or contribute text to the content channels. It is difficult to gage the credentials of the contributors, so the quality of the guides varies. To play it safe, try to back up the material you read on

Internet Resources on Weight Loss and Obesity
© 2007 by The Haworth Press, Inc. All rights reserved.
doi:10.1300/5398_02

About.com with information contained in government and medical organization Web sites.

Dogpile
<http://www.dogpile.com>

Dogpile ("all the best search engines piled into one") is a metasearch engine: it searches several resources simultaneously (Ask Jeeves, Yahoo!, Google, etc.). Dogpile was created in 1996 and is currently owned and operated by InfoSpace, Inc., a wireless and Internet solutions company marketing its services to businesses. Search by specific keywords, and click the Fetch button. Be aware that Dogpile is a commercial entity, accepting advertisements. Some "featured listings" are commercial sites that sell products.

Google
<http://www.google.com>

Google, started in 1998 by former Stanford University graduate students Lawrence Page and Sergey Brin, has become a favorite search engine of medical librarians, academic researchers, and physicians. The phrase "to Google someone" (conduct a Google search for information about a person) has crept into our lexicon. Companies search Google to determine whether prospective employees have lied on their resumes or concealed unfortunate incidents in his or her background. Google ranks Web sites by how frequently other sites link to them, and integrates the content of several smaller search engines. Use Google Groups to locate Usenet discussion groups (Internet forums).

Health on the Net (HON)
<http://www.hon.ch/MedHunt>

HON is a nonprofit medical information gateway produced by Health on the Net Foundation, a Swiss organization. This respected site began in 1995 at a meeting where international telemedicine experts convened to discuss the use of the Internet in healthcare. Attending the meeting were Michael DeBakey, MD, the noted heart surgeon, and officials from the

National Library of Medicine in Baltimore. The attendees voted to "create a permanent body that would . . . promote the effective and reliable use of the new technologies for telemedicine in healthcare around the world." HON now works with the University Hospitals of Geneva and the Swiss Institute of Bioinformatics. The two search engines at this Web site include MedHunt and HONselect. It is considered very prestigious for a medical Web site to bear the HONcode seal of approval. The quality filtering by the HON Foundation is most stringent.

Ixquick Metasearch
\<http://www.ixquick.com\>

Begun in late 1999, Ixquick calls itself the world's most powerful metasearch engine. It searches fourteen search engines at once. Ixquick highlights sponsored (paid) sites by listing them first. Large ads appear on the right side of the page.

Mamma
\<http://www.mamma.com\>

Intasys Corporation now owns Mamma, created in 1996 as a student's masters thesis. Mamma is a metasearch engine that sifts simultaneously through several search engines. This is the search engine to use if you are looking for new, unusual, rare, or obscure topics. Click on the Go Mamma button after typing in a word, name, or phrase. In one respect, Mamma is similar to Google: advertised Web sites are featured ("Mamma Recommends").

MetaCrawler
\<http://www.metacrawler.com\>

MetaCrawler simultaneously searches several of the most popular search engines, including Google, Yahoo!, Ask Jeeves, and Overture. Now owned and operated by InfoSpace, Inc., it was originally developed in 1994 by a graduate student and a professor at the University of Washington. MetaCrawler's motto is "get better results, easier."

Yahoo!
<http://www.yahoo.com>

Yahoo! ("Yet Another Hierarchical Officious Oracle") was created in 1994. Its developers, two engineering students at Stanford University, originally called the search engine "Jerry's Guide to the World Wide Web." Featured Web sites are often ones suggested by users.

ARE YOU THE MASTER OF YOUR DOMAIN?

The short tag at the end of a Web site address (biz, com, edu, gov, mil, net, and org) indicates the origin of the site. The tags com, net, or biz are usually used by businesses, edu is an educational institution such as a school or hospital, mil is used by the military, gov indicates a government (local, state, or national), and org is usually an organization. Sometimes an additional abbreviation follows the domain. This indicates a country or foreign language (fr is France, esp is Spain or Spanish language, ca is Canada, and uk is the United Kingdom). Web sites in the gov or edu domains have the most credibility and the least bias, so choose one of these sites over ones displaying other domains. Warning: carefully note the domain designation when purchasing products over the Internet. Never respond to unsolicited e-mail messages directing you to vaguely familiar Web sites. Sometimes fraudulent sites have nearly the same URLs and logos as reputable sites, but the domains or country designations are different. Look for a locked padlock on the Web site and the "https" secure site designation before making a purchase.

TOO GOOD TO BE TRUE:
HOW TO SORT THE FAKES FROM THE DIAMONDS

Have you been playing cyber roulette when seeking diet information? Would you take medical advice from a high school student? Unless you keep quality guidelines in mind when searching the Internet, you might find yourself reading an attractive Web site that contains slanted, incorrect, or even dangerously outdated information. With easily purchased or downloaded software, anyone can create a professional-looking Web site that appears to be authoritative. How do you determine what is authentic and credible? How do you spot fraud? The following characteristics should be evident on authoritative sites:

- *The identity of the Web site creator:* Is the person or group responsible for the content readily identified? Can you contact them by e-mail, fax, mail, or telephone? What are the credentials of the creator (organizational affiliation, education/training, experience, and publications)?
- *Currency:* When was the site created? Was the site updated recently? Are the links to other Web sites still valid (do they work)? Is the site frequently "down" for maintenance?
- *Seal of approval:* Has the federal government or a national health organization recommended the site? Does it bear the "HONcode" logo, the seal of approval awarded to sites that meet the HONcode of Conduct? *See* **Health on the Net Foundation (HON) (http://www. hon.ch)**. HON, created in 1995, is an international Swiss organization whose mission is to guide Internet searchers to authoritative online health and medical information. HON also sets stringent ethical standards for those who develop health and medical Web sites. HON categorizes its selected Web sites as educational, individual, or commercial. Another organization accrediting health Web sites is **URAC (http://www.urac.org/)**. "Promoting quality healthcare" is the motto of URAC. This nonprofit charitable organization, also known as American Accreditation Healthcare Commission, was founded in 1990. Its goal is to establish standards for the healthcare industry. URAC's mission is "to promote continuous improvement in the quality and efficiency of healthcare delivery by achieving a common understanding of excellence . . . through the establishment of standards . . . and a process of accreditation." The consumer section of the URAC site lists and links to (as of February 2006) approximately 300 sites that have received URAC accreditation. Under URAC's Health Web Site Accreditation Program, sites are evaluated for disclosure of financial backing and sponsorship, privacy and security, and quality and oversight standards.
- *Credibility and validity:* For a guide to determining a medical Web site's credibility and to view fraud warnings, visit **Quackwatch (http://www.quackwatch.com)**. Another site exposing medical

fraud is the **National Council Against Health Fraud (http://www. ncahf.org)**. NCAHF is an excellent source for viewing the latest news on medical fraud. The articles that appear on NCAHF are written by health professionals.

- *Bias:* Is the site selling books, vitamins, and devices? Is it sponsored or produced by a drug company? Does the site advertise a practitioner's clinic? What is the hidden agenda?
- *Purpose:* Is the mission or are the objectives evident? Is the purpose of the site to educate, inform, support, sell a product, or attract customers or patients?
- *Audience:* Who is the intended audience: health professionals or consumers?
- *Attractiveness or visual appeal:* Is the Web site easy to navigate and exciting to view? Are there good illustrations and tables or bullets highlighting key points?
- *Origin:* What is the Web site domain? Domains tagged gov and edu have the most credibility. Do you know the country of origin of the Web site?
- *Content:* Does it provide the information you seek and with sufficient depth? Is the information accurate and current? What is the source of the information: opinion, observation, research, journal articles, or books? How does the content compare with information in similar resources?

Chapter 3

The Internet or the World Wide Web: Are the Terms Interchangeable?

The terms Internet and World Wide Web (or simply, the Web) are often used interchangeably. In fact, the Web is only a portion of the Internet. Some computer users spend most of their online time exchanging opinions in chat rooms or reading material in newsgroups and mailing lists. They seldom look at Web sites.

DISCUSSION GROUPS: CHAT ROOMS, MAILING LISTS, AND NEWSGROUPS

For good advice on distinguishing a mailing list (moderated or unmoderated) from a newsgroup, Usenet group, chat room, forum, bulletin board, or online support group, read Google Groups—Basics of Usenet (http://groups.google.com/googlegroups/basics.html); for definitions and links to other resources, read Mossresourcenet (http://www.mossresourcenet.org).

Users "subscribe" or register to send and receive messages on a moderated mailing list, online support group, or forum. Sometimes the "list owner" or moderator establishes rules for posting messages and participating in discussions. Usually, the moderator specifies categories of approved discussion topics. Some groups are "chatty" or "high traffic" and generate dozens of messages each day.

Internet Resources on Weight Loss and Obesity
© 2007 by The Haworth Press, Inc. All rights reserved.
doi:10.1300/5398_03

With most chat rooms, bulletin boards, and Usenet groups, there is no need to subscribe. You simply access the group's Internet site and scan current and archived (old) messages and decide whether you want to join in the message threads. To locate groups on dieting or weight loss, go to the following sites and type in terms such as diets, dieting, obesity, weight loss, or overweight.

Google Groups
<www.google.com>

Select the section "Groups" and then type a word or phrase. Google Groups lists Usenet (newsgroups) discussion forums. The site has an excellent FAQs section that answers a beginner's questions about searching and using newsgroups.

OneList.com
See **Yahoo! Groups**

SupportPath.com
<www.supportpath.com>

This Web site provides links to support-related bulletin boards, chat rooms, local and national support organizations, meetings, and information on dozens of health problems. Formerly called Support-Group.com, its philosophy is that Internet discussion groups offer participants the opportunity to benefit from shared experiences (and not feel isolated), to express anger, fear, and sadness, and to gain hope. Participants often share personal stories.

Yahoo! Groups (formerly OneList.com)
<groups.yahoo.com>

This is part of Yahoo.com. Readers can create their own discussion groups. The service is free, and there are easy-to-follow instructions. Registration is required. Also, use Yahoo! Groups to search for existing online groups.

SAGE ADVICE FROM CYBER AUNT:
DO YOU WANT TO BARE ALL ON THE INTERNET?

Warning: Discussion groups may save messages (archive) for public access or subscriber member access, so do not include confidential or personal information in your postings. Do not badmouth your boss or reveal company secrets. See the following section for additional advice about participating in online discussion groups.

Caution: It is difficult to verify the credentials (background, authority, honesty, or knowledge) of contributors to online discussion forums. Never follow recommendations offered by participants until you have checked with your dietitian or healthcare provider (even if the person in the chat room says he or she is a physician or expert). Please be cautious.

MANNERS COUNT:
PRACTICE GOOD NETIQUETTE

Because support groups are so friendly and informal, it is tempting to act on impulse and dash off a message, forgetting basic rules of etiquette, grammar, and spelling. Also, some features are unique to the Internet. Here is some advice:

- Do not reveal personal or confidential information when posting a message to a group (whether a mailing list, bulletin board, chat room, or online support group). Messages may be stored (archived) indefinitely or made available to the public.
- Do not forward messages from other people without first obtaining their permission.
- Do not type in ALL CAPITAL LETTERS. This is annoying and is known as "shouting."
- Do not send replies in a fit of anger.
- Do not insult ("flame") someone over the Internet.
- Do not stalk or harass others in cyberspace.
- Carefully consider the impression your message leaves. Would your words be perceived as insulting or insensitive?
- Do not fill up mailboxes by posting "me too" messages. As a rule, do not reply or post a message unless you have a new or unique opinion to add.
- Reply to individuals, rather than to the entire group, when appropriate.

- There is no shame in stating that you are a "newbie" (newcomer) to discussion lists.
- Include a subject line when sending messages. Blank subject lines, especially on chatty (high traffic) discussion groups, waste members' time. Some moderated discussion groups require participants to use agreed-upon subject categories so that participants can delete messages that do not interest them.
- Do not be afraid to "lurk" for a while (read, but not post messages) when you first join a discussion group. It takes a while to get a feel for the "culture" of a group. Do not overuse emoticons (the punctuation marks used to express emotion in e-mail) or acronyms (abbreviations in which each letter stands for initials of each word). Liberal use of emoticons and elaborate signature files has become *passé*. Ten years ago, when e-mail was still a new communication mode, it was considered clever to use emoticons, acronyms, smilies, and cute signatures ("sig files") with portraits of one's dog, exploding firecrackers, or the Golden Gate Bridge, composed entirely of keyboard characters. Now, it looks silly if they are overused or appear in business communications. Examples of emoticons and acronyms follow:

:-X	my lips are sealed
[]	hug
:(	sad
:)	happy
lol	laughing out loud
rotfl	rolling on the floor laughing
rtfm	read the _____ manual!
btw	by the way
THX	thanks
%-)	confused
FAQ	frequently asked question

- For additional emoticons and abbreviations, see Emoticons & Smilies Page at www.muller-godschalk.com/emoticon.html.
- Follow the rules at work. Most companies have explicit guidelines and policies about appropriate use of the Internet and e-mail. Most companies monitor employees' messages and Internet activities. When in doubt, use your home computer and a personal e-mail address. It would be stupid to lose your job for violating your company Internet or e-mail policies.

Chapter 4

General Diet and Nutrition Web Sites: Nutrition 101

No matter which diet you follow or what your general physical condition is, dietitians and healthcare providers suggest you first educate yourself about the fundamentals of good nutrition. The following Web sites are excellent resources for information on dieting and nutrition. Most feature statistics, calculators, lists, and links to additional sites.

American Academy of Pediatrics
<www.aap.org>

The American Academy of Pediatrics (AAP) site offers comprehensive information on children's nutrition and obesity. Select "Children's Health Topics," browse topics in alphabetical directories, and choose "Nutrition." Readers can purchase *Guide to Your Child's Nutrition* and *Pediatric Nutrition Handbook,* but these books are expensive. Pamphlets on related topics are for sale, in bulk, to pediatricians, who may offer single copies to patients at no charge: *Feeding Kids Right Isn't Always Easy* and *Starting Solid Foods.* Free, downloadable materials, including documents on breastfeeding, are listed in the section "AAP Resources."

American Dietetic Association/Eatright.org
<www.eatright.org>

This is a great Web site to bookmark because it is authoritative and thorough. You will want to refer to it often. The American Dietetic Association

Internet Resources on Weight Loss and Obesity
© 2007 by The Haworth Press, Inc. All rights reserved.
doi:10.1300/5398_04

(ADA) is the national association of food and nutrition professionals, with over 70,000 members. The association's mission is to promote health, well-being, and "optimal" nutrition for the public. The site includes an archive of dozens of "Tips of the Day" that cover nutrition topics of interest to the public (vegetarianism, BMI as a gage of obesity, antioxidants, juices, iron, etc.). Type your zip code, and the ADA lists nutrition professionals in your town. The association explains the educational credentials of registered dietitians and registered dietetic technicians. The site features healthy recipes, including back-to-school lunches. Read "Nutrition Fact Sheets," "Special Needs," and "Weight Management." Do not overlook "Ten Red Flags That Signal Bad Nutrition Advice." Study the illustration and detailed explanation of the US Department of Food and Agriculture's new food guide pyramid. The ADA site also provides a link to the government's separate food guide pyramid for children. The "Product Catalog" sells books and book packages (money-saving collections of materials on related subjects such as pediatric nutrition and sports nutrition).

Ask the Dietitian
<www.dietitian.com>

Joanne Larsen, MS, a registered and licensed dietitian, created and maintains this Web site. There is an extensive section consisting of answers to nutrition-related questions posted by the public. For each question, Larsen gives a thorough answer with detailed recommendations. "Top Ten Tips to Spot Nutrition Quackery" and "Ten Changes You Can Make to Lose Weight" are two featured sections. Larsen has written books, nutrition software, and databases, and she appears on television and radio. Ask the Dietitian demonstrates her direct approach to nutrition advice. This is a great source for personalized and free nutrition information.

The Blonz Guide to Nutrition, Food and Health Resources
<www.blonz.com>

Edward Blonz holds MS and PhD degrees in nutrition from the University of California at Davis. He is a member of the Dietary Supplement Advisory Council of the U.S. Food and Drug Administration. Blonz is a certified nutrition specialist who appears on national television and writes a newspaper column and books. His Web site has won many awards. Blonz's site includes links to nutrition resources containing "science-reliable" evidence. He groups the links into categories: government; food resources and associations (scores of Web sites from Taco Bell to Godiva Chocolate Web sites are linked here); nutrition, food, and fitness; search engines; and academic institutions having food and nutrition programs. Offbeat, but authoritative!

Calorie Control Council
<www.caloriecontrol.org>

The Calorie Control Council was established in 1966. It is an international, nonprofit organization representing low-fat and reduced-calorie food and beverage manufacturers and suppliers. The site features recipes, a free template of a daily food diary, online calorie calculators, a BMI calculator, a healthy-weight calculator, and general information on cutting calories and fat. Hot topics are healthy low-carb dieting and current, major news stories about diet and nutrition.

Center for Nutrition Policy and Promotion
<www.cnpp.usda.gov>

Center for Nutrition Policy and Promotion's (CNPP) objective is to improve "the nutrition and well-being of Americans." This is the home of the controversial new food pyramid (MyPyramid) and "Dietary Guidelines for Americans." CNPP is part of the US Department of Agriculture (USDA). CNPP's mission is to link scientific research to the nutrition needs of American consumers and to translate the research into information and educational materials for policymakers, consumers, and health, education, and industry professionals (Figure 4.1).

Center for Science in the Public Interest
<www.cspinet.org>

Select "Nutrition and Health" to access the "Health, Nutrition, and Diet" page with calculators, lists of ten foods you should or should never eat, and ten steps to a healthy diet. The site emphasizes information about food safety and good nutrition. Center for Science in the Public Interest (CSPI) championed the elimination of trans fats from foods years before the 2003 government requirement that Nutrition Facts labels disclose trans fat content. Read selected articles from CSPI's *Nutrition Action Healthletter.*

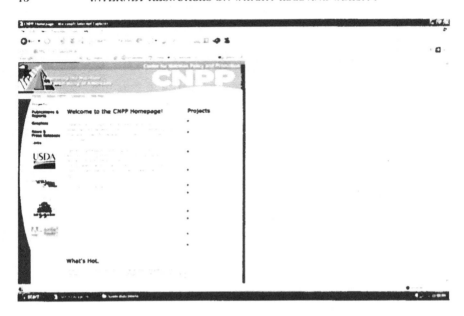

FIGURE 4.1. Center for Nutrition Policy and Promotion
<www.cnpp.usda.gov>

Food and Nutrition Information Center
<www.nal.usda.gov/fnic>

This is the place to go to find the USDA MyPyramid, formerly called the Food Guide Pyramid. The pyramid was revised and renamed in January 2005. The public was invited to submit ideas and comments before the release of the new guide, but many consumers and experts find the new pyramid confusing and the guidelines difficult to follow. See www.mypyramid.gov for more details about using the interactive MyPyramid. The Food and Nutrition Information Center (FNIC) is part of the National Agricultural Library and has been providing nutrition information to consumers, teachers, government employees, and health professionals since 1971. At the FNIC site, read research reports, fact sheets, government reports, and educational materials. The 2005 sixth edition of *Dietary Guidelines for Americans* is available in full text here. The Web site includes several related pamphlets, designed to help Americans put the guidelines into practice. Notice the list of downloadable food guidelines from around the world; for example, there are guidelines from Canada, Norway, Greece, Singapore, and Great Britain. Other sections include "Dietary Supplements and Herbal Information," "Alternative Medicine," and a database listing the nutrient composition of over 6,600 foods. FNIC packs a lot of useful, science-based information into its Web site. Although the new pyramid has invoked much criticism, in an area rife with quackery and false claims, this government resource is outstanding for its common sense and integrity.

Gateway to Government Nutrition Sites/Nutrition.gov
<www.nutrition.gov>

The gateway gathers all government nutrition resources in one site and attempts to fulfill President George W. Bush's HealthierUS mandate by promoting physical fitness, nutrition, prevention (health screening), and avoidance of risky behaviors. The section "Shopping, Cooking, and Meal Planning" consolidates information on meal planning, shopping, food storage, food preservation, food labeling, and cooking methods. There is also a section on recipes, including cooking for children, and healthy African-American and Latino recipes.

International Food Information Council
<www.ific.org>

The International Food Information Council (IFIC) is "your nutrition and food safety source." The site is available in Spanish as well as English. IFIC, through the IFIC Foundation, provides science-based information to government officials, nutrition and health professionals, and journalists who educate the public about nutrition and food safety. The foundation aims to "bridge the gap" between science and communications. IFIC is supported by the food, agricultural, and beverage industries. There is a glossary of food-related terms. Through two separate sections "Go to Food Safety Information" and "Go to Nutrition Information" the IFIC Web site provides a wealth of reliable, current reference information. Teachers, especially, will find these resources valuable.

MyPyramid.gov
<www.mypyramid.gov>

The widely touted 2005 recasting of the Food Pyramid was released with much fanfare. The online-only interactive MyPyramid.gov allows individuals to personalize diet recommendations based on activity level, age, and gender. So many visitors tried to access the site on the first day that it crashed. Critics contend that the required Internet access excludes some low-income consumers, that MyPyramid.gov places too much emphasis on dairy foods, and that the personalization does not take into account height and weight. At one point, MyPyramid.gov was even part of the story line of a popular syndicated newspaper comic strip. There is a separate food pyramid for children. See additional information under Food and Nutrition Information Center (Figure 4.2).

National Academy of Sciences/Food and Nutrition Board
<www.iom.edu/boards.asp>

The Food and Nutrition Board is part of the Institute of Medicine, one of the National Academies of Science. The board was established in 1940, and its mission is to study issues related to the adequacy and safety of the U.S. food supply, establish guidelines for "adequate" nutrition, and render judgments on the relationship between health, nutrition, and food intake. Unlike FNIC, it does not answer questions from consumers or serve as a nutrition reference resource. It sponsors symposia and conducts research projects. Current projects include research on how food marketing influences the diet of children and prevention of childhood obesity.

National Dairy Council
<www.nationaldairycouncil.org>

The National Dairy Council (NDC) is the nutrition-marketing arm of Dairy Management, Inc., which produces the 3-A-Day Dairy program with the American Dairy Council. NDC has been in existence since 1915. Through its nutritionists and dietitians, NDC provides science-based nutrition information to health professionals, teachers, consumers, and the media. Through its network of state and regional dairy councils, it educates schoolchildren. The Web site includes recipes and health tips. NDC's pro-

FIGURE 4.2. MyPyramid.gov
<www.mypyramid.gov>

gram "Healthy Weight with Dairy" is designed around recent research studies that claim a link between consumption of three to four servings a day of dairy products and more effective weight loss, less abdominal fat, and lower incidence of kidney stones. (Note that the results of these studies are being questioned by other nutrition organizations.) The site includes a BMI calculator, a link to the new federal Food Guidance System, including MyPyramid, and a database of nutritional values (calories and nutrients in foods). *See also* **3-A-Day.org** (www.3aday.org).

National Institute of Nutrition
<www.nin.ca>

National Institute of Nutrition (NIN) is a Canadian organization bringing together nutritional science with government, industry, and consumers to produce communication programs and research projects. Its mission is to serve as the "catalyst for advancing the nutritional health of Canadians." The Web site is available in French and English. A straightforward Web site with few illustrations and dense text, NIN displays topics on the right-hand side of the home page (Canada Food Guide, diabetes, dietary fat, vegetarian eating, nutrition policy, and more). In the FAQ section, there are links for locating registered dietitians in Canada ("Find-a-Dietitian" and "Dial-a-Dietitian"). The site includes full-text articles from nutrition journals published by NIN.

Nutrition and Your Health:
  Dietary Guidelines for Americans
<www.nal.usda.gov/fnic/dga/index.html>

See **Food and Nutrition Information Center** (www.nal.usda.gov/fnic) in this chapter.

Nutrition Fact Sheets
<www.eatright.org/public>

There is a wealth of useful information in this section of the American Dietetic Association's Web site. Some of the free fact sheets, download-able in pdf format, include The Power of Potatoes, The Pasta Meal, Lycopene, Canned Food, Beef, Shopping Solutions, and African-American Health and Dairy Foods. When it comes to nutrition, dietitians are the experts.

Oldways
<www.oldwayspt.org>

Oldways Preservation and Exchange Trust, a "food issues think tank," is a Boston-based, nonprofit organization founded in 1990 by K. Dun Gifford, a Harvard-educated attorney who has owned and operated several restaurants. Gifford has worked with Senator Edward Kennedy and the late Senator Robert Kennedy. He is the former national chair of the American Institute of Wine & Food. Oldways combines nutritional science with information on fine cuisine. It educates consumers, government officials, farmers, members of the food industry, and chefs. Oldways believes that traditional foods and eating patterns are healthier than modern Western ways of eating. It promotes symposia, conferences, and tours. Recently, in collaboration with the Harvard University School of Public Health, it offered weeklong continuing education programs in Italy for physicians, dietitians, and allied health professionals wanting to study the Mediterranean diet. With input from 500 scientists, Oldways developed five eating pyramids: Asian, Latin American, Mediterranean, and Vegetarian, plus a fusion called Eatwise Food Pyramid. To download the healthy eating pyramids, select "Wise Eating" and then "Traditional Diet Pyramids." Oldways' Eatwise program is designed to help families eat less junk food and follow healthier eating habits. The program emphasizes exercise and whole grains. It recognizes, however, that people need to accommodate eating frequent, abundant or "feast" meals in their lives. Feasts include meals consumed at parties, picnics, and restaurants. Oldways developed a cooking and nutrition program for schoolchildren. This site also features recipes.

Physicians Committee for Responsible Medicine
<www.pcrm.org>

Physicians Committee for Responsible Medicine (PCRM) is a nonprofit organization of physicians and nonprofessionals, founded in 1985. The current president is Neal D. Barnard, MD. Barnard has published several books, and frequently lectures to groups of physicians. He is a dynamic speaker and makes his point with a reasonable, nonstrident manner. PCRM's advisory board includes physicians, nutritionists, and scientists. The organization promotes preventive medicine, with an emphasis on healthy nutrition with a vegetarian slant. PCRM introduced the "New Four Food Groups": fruits, legumes (peas, beans, and lentils), whole grains, and vegetables. In 2005, they began a public service announcement campaign disputing the National Dairy Council and several dairy food manufacturers' linking of dairy food consumption with weight loss. In 2004, PCRM launched a vigorous print and television campaign against low-carb, high-protein diets, specifically the Atkins diet program. They maintain a registry to document adverse effects of such diets. Current initiatives include the Healthy School Lunch campaign, alternatives to animal dissection in health science education, and the Cancer Project, which studies prevention of cancer and increasing survival time for cancer patients by means of better nutrition. PCRM offers vegetarian cooking classes for cancer survivors. Download the textbook for the classes, *The Survivor's Handbook*, from the PCRM Web site. The Strong Bones campaign disputes the theory that milk is essential for strong bones ("Milk: It's Not All It's Cracked Up To Be"). The PCRM Web site includes fact sheets on nutrition topics, including *Guide to Healthier Weight Loss,* a three-week low-fat, and vegetarian diet plan.

To download "Vegetarian Starter Kit," select "Vegetarian Diet" after selecting "Health" from the home page. There are additional vegetarian materials here, too.

3-A-Day.org—Dairy Information for Families
<www.3aday.org>

Dairy Management, Inc., of the American Dairy Association, manages this resource. The site includes recipes, news items, and advice on nutrition. *Get 3!* is a free online newsletter providing recipes, news items, nutrition information, and advice from registered dietitians. For health professionals, the site contains digests of research studies reported in medical and nutrition journals. The 3-A-Day of Dairy campaign receives support from the American Academy of Family Physicians, American Academy of Pediatrics, National Medical Association, and American Dietetic Association. Note that medical experts have recently cast doubt on the quality and quantity of research studies linking consumption of dairy products with weight loss.[1]

U.S. Department of Agriculture National Nutrient Database
from Nutrition Data Laboratory
<www.nal.usda.gov/fnic/foodcomp/index.html>

The databases on this site are comprehensive resources providing information about the nutrient content of foods. Using the Nutrient Lists database, type "calcium," for example, and view a list of the calcium content of hundreds of foods. Search the USDA National Nutrient Database (NND) for Standard Reference; for "apricot" as an example, choose "apricots, raw" and view a detailed listing of the nutrients in one raw apricot.

Wheat Foods Council: Grain Nutrition Information Center
<www.wheatfoods.org>

This national nonprofit organization maintains the Grain Nutrition Information Center at this site. With the publication of MyPyramid, there has been renewed emphasis on consuming whole grains as part of a healthy diet. The mission of the Wheat Foods Council is to increase the public awareness of the healthful qualities of this food group. The site includes recipes.

NOTE

1. Bowen, J., Noakes, M., and Clifton, P.M. (2005). Effect of calcium and dairy foods in high protein, energy-restricted diets on weight loss and metabolic parameters in overweight adults. *International Journal of Obesity* (London) 29:957-965.

Chapter 5

Health and Dietary Assessment
Web Sites

Put down the handheld calculator, forget the rusty math skills, and lose the pen and paper. The resources in the following Web sites simplify calculation of body mass index (BMI) and calorie expenditure, and explain the information on food labels. Also, see Chapter 4 for additional information about calories, nutrient content of foods, and BMI.

ONLINE CALCULATORS
AND ASSESSMENT TOOLS

Basal Metabolism Calculator
<www.room42.com/nutrition/basal.shtml>

Type your height, weight, age, sex, and activity level, and the Room42. com software will let you know if you are overweight, obese, underweight, or just right. As the developer of this algorithm used for the calculations is not a medical professional, she prudently includes links to the United States Department of Agriculture (USDA) Dietary Guidelines Height and Weight Tables.

Calorie Control Council
<www.caloriecontrol.org>

There are a number of calculators at this site: BMI calculator, exercise calculator. two calorie calculators, weight-maintenance calculator, and

Internet Resources on Weight Loss and Obesity
© 2007 by The Haworth Press, Inc. All rights reserved.
doi:10.1300/5398_05

healthy-weight calculator. There is a recipe section, articles on weight-loss topics, and digests of current news on weight issues and health (see Figure 5.1).

Light 'n Fit
<www.lightnfit.com>

As part of its weight management program, Light 'n Fit placed a BMI calculator and a calcium calculator on its Web site. The site also features exercise and general weight-loss tips and aids. The Light 'n Fit Right Fit Plan for weight management incorporates sensible eating, featuring yogurt

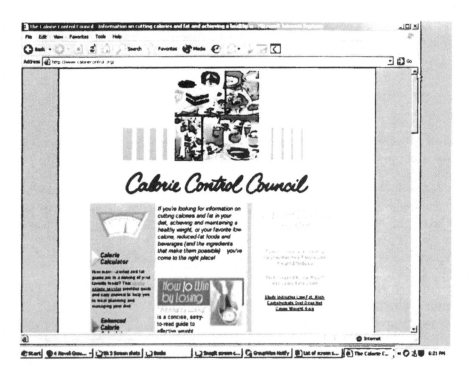

FIGURE 5.1. Calorie Control Council
<www.caloriecontrol.org>
Reprinted with permission of Calorie Control Council.

and other high-calcium foods, and moderate exercise. Carolyn O'Neil, a prominent registered dietitian, developed the four-week diet plan in collaboration with Light 'n Fit.

Nutrition Analysis Tools and System's
Energy Calculator
<nat.crgq.com/energy/daily.html>

Chris Hewes, a computer programmer, and Jim Painter, a registered dietitian, from the Department of Food Science at the University of Illinois, developed the Nutrition Analysis Tool (NAT) software in 1996. The energy calculator takes input of age, sex, height, weight, and amount of time in various levels of activity during a twenty-four-hour period, and calculates the number of calories expended.

RealAge Nutrition Health Assessment
<www.realage.com>

RealAge is a consumer health media company. The Web site and the health assessment tools are sponsored by several pharmaceutical and diet program companies, so there is a strong commercial slant to RealAge. Fill out the online questionnaire and see how closely your chronological age matches your biological age. Follow additional links for in-depth health and nutrition assessments. RealAge offers good advice and positive encouragement.

Shape Up America! Pediatric BMI
Assessment Tool
<www.shapeup.org/sua>

In 1994, former U.S. Surgeon General C. Everett Koop, MD. launched Shape Up America! (SUA), a nonprofit organization. SUA's objective is to find solutions to the problem of overweight in America. Through a collaborative effort of experts and organizations in the fields of health, nutrition, and physical fitness, SUA promotes healthy weight and increased physical activity. The Pediatric BMI Assessment Tool is a good resource to aid parents and health professionals dealing with the epidemic of obesity in children.

UNDERSTANDING FOOD LABELS

Since 1990, when the Nutrition Labeling and Education Act called for overhauling of food labels, the FDA and the USDA have worked to develop informative, easier-to-read food labels using standardized terminology. The new labels were put into use in 1994. Recently, information on trans fat and food allergen content was added to food labels. Labeling information helps consumers calculate diet points (Weight Watchers dieters), food exchanges (diabetics), fats (low-fat dieters, Weight Watchers dieters, and heart disease patients), carbohydrates (Atkins dieters, Zone dieters, South Beach dieters, and diabetics), sodium (kidney disease patients, heart disease patients, and hypertension patients), protein (Zone dieters, kidney disease patients), and calcium (people diagnosed with osteoporosis).

**Food and Drug Administration/Center
for Food Safety and Applied Nutrition:
How to Understand and Use the Nutrition Facts Label
<www.cfsan.fda.gov/~dms/foodlab.html>**

Take the comprehensive tutorial on reading and understanding food labels. There is a video segment and supplementary downloadable documents.

**KidsHealth for Parents: Deciphering Food Labels
<www.kidshealth.org/parent>**

Select the "Nutrition & Fitness" section and then "Deciphering Food Labels." KidsHealth, from the Nemours Foundation's Center for Children's Health Media, is an award-winning Web site. The section on food labels describes the history of the new food-labeling system and features a comprehensive explanation of the components of the food label. KidsHealth also features recipes and materials written for children and teens.

MayoClinic.com. Reading Food Labels
<www.mayoclinic.com>

Select "Healthy Living" and then "Food & Nutrition Center" to read material on food labeling, including an interactive instructional session on how to read food labels.

Chapter 6

Eating Disorders

ANOREXIA NERVOSA AND BULIMIA NERVOSA

Academy for Eating Disorders
<www.aedweb.org>

The Academy for Eating Disorders (AED) is a multidisciplinary, international group of medical specialists with interest in anorexia nervosa, bulimia nervosa, binge eating, and related disorders. The mission of the organization is to promote excellence in research, treatment, and prevention of eating disorders and to disseminate information regarding these to the general public. AED is also an advocate on behalf of patients, the public, and eating-disorder professionals.

Anorexia Nervosa and Related Eating Disorders
<www.anred.com>

Anorexia Nervosa and Related Eating Disorders (ANRED), a nonprofit organization run by healthcare professionals, provides statistics and other information about eating disorders, as they occur in both men and women and in all age groups. Print documents on over fifty topics related to obesity, anorexia, bulimia, binge eating, warning signs, special risk factors, causes, medical and psychological complications, treatment, and recovery.

Internet Resources on Weight Loss and Obesity
© 2007 by The Haworth Press, Inc. All rights reserved.
doi:10.1300/5398_06

Bulimia Nervosa Resource Guide
<www.bulimiaguide.org>

Bulimia Nervosa Resource Guide, launched in February 2006, is produced by ECRI, a health services research agency located in suburban Philadelphia. The research project, funded by a $300,000 grant from a private foundation, culminated in the Web site. The site covers how to recognize the problem in loved ones and how to evaluate treatment options and maximize insurance coverage. This site provides checklists and tips to help families negotiate the tangle of red tape in dealing with diagnosis and treatment. Read the report *Bulimia Nervosa: Efficacy of Available Treatments* (see Figure 6.1).

A Chance to Heal
<www.achancetoheal.org>

A Chance to Heal is a nonprofit organization started in 2004 by a mother whose daughter has bulimia. The organization raises funds to increase

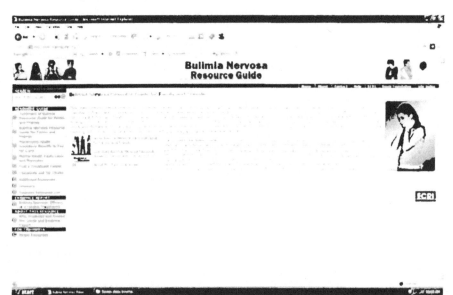

FIGURE 6.1. Bulimia Nervosa Resource Guide
<www.bulimiaguide.org>
Reprinted with permission of Bulimia Nervosa Resource Guide.

awareness about the disorder. advocate for sufferers, and assist those in need of treatment. The foundation raised over $100,000 in its first year. Even families with medical insurance may find they need to pay out of pocket for adequate treatment. A Chance to Heal helps these families, too (see Figure 6.2).

National Eating Disorders Association
<www.nationaleatingdisorders.org>

National Eating Disorders Association (NEDA) is a large, U.S.-based nonprofit organization covering all types of eating disorders. The Web site is comprehensive, including sections on research grants, volunteer and internship opportunities, support groups, treatment referral, and educational

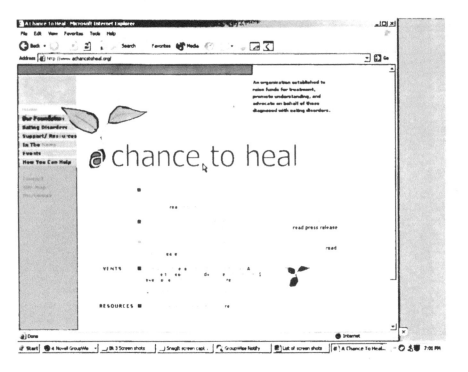

FIGUFE 6.2. A Chance to Heal
<www.achancetoheal.org>
Reprinted with permission of A Chance to Heal.

opportunities. NEDA sponsors National Eating Disorders Awareness Week and has a celebrity "ambassador," Paula Abdul (see Figure 6.3).

COMPULSIVE OVEREATING (BINGE EATING)

Overeaters Anonymous
<www.oa.org>

Overeaters Anonymous (OA) offers a twelve-step recovery program for compulsive overeaters. There are groups meeting all over the world. Members pay no dues or fees, but assist new members and at meetings. The meetings and other tools provide a fellowship of experience, hope, and strength in which members respect each other's anonymity. The OA Web site has a meeting locator. Some groups meet on Sundays. Read the

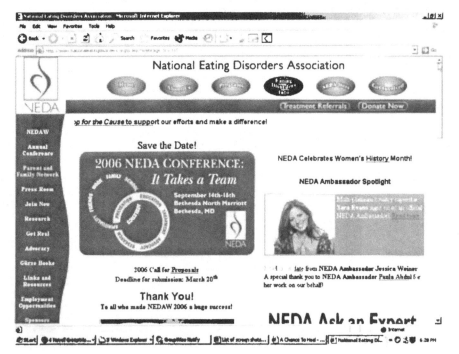

FIGURE 6.3. National Eating Disorders Association
<www.nationaleatingdisorders.org>
Reprinted with permission of the National Eating Disorders Association.

"Twelve Concepts of Service" that form the structure of OA. The "Twelve Steps" and "Twelve Traditions" were adapted from Alcoholics Anonymous. According to Naomi Lippel, managing director of OA, the meetings address the physical, spiritual, and emotional well-being of participants. It is not a religious or political organization.[1]

NOTE

1. Personal communication from Naomi Lippel, managing director, Overeaters Anonymous, Inc., December 27, 2005.

Chapter 7

Obesity in Children and Teenagers

American Academy of Family Physicians: Familydoctor.org
<familydoctor.org/343.xml>

Familydoctor.org is aimed at the general public. Obesity and Children: Helping Your Child Lose Weight is one of its sections. The format is Q&A, but additional information is provided through a link to a section on food and nutrition. Some material is available in Spanish.

American Academy of Pediatrics
<www.aap.org>

Read American Academy of Pediatrics (AAP) recommendations on the prevention of pediatric overweight and obesity, view the AAP Obesity and Overweight Web site, and download the full text of the AAP Policy Statement on Prevention of Pediatric Overweight and Obesity.

Centers for Disease Control:
The Biggest Generation, Connect with Kids
<www.cdc.gov/healthyyouth/healthtopics/connect.htm>

The Biggest Generation is the title of a television program to be broadcast in 2006 on major network television stations, which have collaborated with Florida's Department of Health and the Centers for Disease Control's (CDC's) Division of Adolescent and School Health and Division of Nutrition and Physical Activity. The name of the partnership is Connect with Kids. Parents and teachers can download materials to be used in conjunction with the television program. The Web site includes links to CDC and

Internet Resources on Weight Loss and Obesity
© 2007 by The Haworth Press, Inc. All rights reserved.
doi:10.1300/5398_07

other government agency resources on dealing with the epidemic of child-hood and adolescent obesity, as well as obesity in adults.

KidsHealth
<kidshealth.org>

The URAC-accredited Web site has teamed ("Educational Partners") with several medical associations, hospital networks, government medical agencies, and children's hospitals to offer current, physician-reviewed information. Select "Teens Site" to view articles such as "Figuring Out Fat," "Why You Need More Than a Scale," and "Food & Fitness." There is a BMI calculator, recipes, and a section on "What's the Right Weight for My Height?" There are separate sections for children and parents. The material is also available in Spanish. Both the children's and teens' sections are written in age-appropriate language. KidsHealth has so much quality health information that it should be bookmarked on every parent's computer (see Figure 7.1).

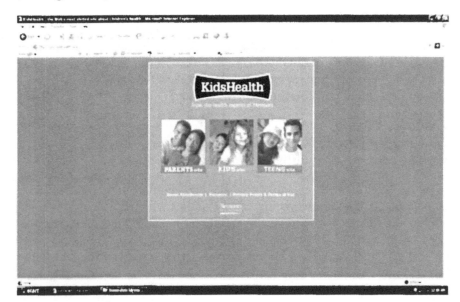

FIGURE 7.1. KidsHealth
<kidshealth.org>
Reprinted with permission of KidsHealth.

National Childhood Obesity Foundation
<www.ncof.org>

National Childhood Obesity Foundation (NCOF) is a nonprofit organization that takes a proactive stance to prevent as well as deal with the existing problem of obesity in toddlers, children, and teens. The organization's board of advisors includes physicians, educators, psychologists, and registered dietitians. Their proposed programs emphasize increased physical activity and better nutrition. NCOF views parents and teachers as the gatekeepers of children's health and nutrition because they have the most control over children's eating habits and activities. NCOF was incorporated in September 2004, but it is awaiting IRS approval before it can solicit funds. Once the foundation has achieved some fund-raising goals, it plans to offer subscribers *NCOF P.O.W.E.R. Healthcare Newsletter,* which will be available in print or electronic format and will be free of cost. Proposed programming includes T.H.I.N.K. (Teaching Health in Nutrition and Kinesiology) and P.O.W.E.R. (Pan-American Obesity Walk for Education and Research) (see Figure 7.2).

National Guideline Clearinghouse: Overweight in Children and Adolescents: Pathophysiology, Consequences, Prevention, and Treatment
<www.guidelines.gov>

The National Guideline Clearinghouse gathers guidelines from national organizations and government Web sites. The resource on overweight in children and adolescents is directed to physicians, especially family physicians. The guidelines were originally developed by the American Heart Association and appeared in the medical journal *Circulation* in April 2005.[1]

Obesity in Children (on MEDLINEplus)
<www.medlineplus.gov>

MEDLINEplus, produced by the National Library of Medicine, is a Web site for consumers. All links are to credible sources, including national organizations, government agencies, major hospitals, and medical schools.

FIGURE 7.2. National Childhood Obesity Foundation
<www.ncof.org>
Reprinted with permission of National Childhood Obesity Foundation.

Read the sections offering the latest news about obesity in children, symptoms and diagnosis, treatment, screening and prevention, statistics, research, a glossary, a directory of nutrition professionals, and organizations. From the A-to-Z list of health topics, select "O" and then "Obesity in Children."

NOTE

1. Daniels, S.R., Arnett, D.K., Eckel, R.H., Gidding, S.S., et al. (2005). Overweight in children and adolescents: Pathophysiology, consequences, prevention, and treatment. *Circulation* 111:1999-2012.

Chapter 8

Weight-Loss Web Sites (Nonsurgical)

A study in *Archives of Pediatrics and Adolescent Medicine* in 2004[1] compared adolescents in thirteen European countries, Israel, and the United States. The United States had the highest prevalence of overweight adolescents. The study based its findings on body mass index (BMI) calculation rather than height/weight charts.

Only about one-third of American adults are of normal weight (BMIs below 25). Obese adults in the United States account for 36 percent higher average annual medical expenditures compared with those of normal weight.[2] The nation spent $75.1 billion in 2003 on drugs, doctor visits, and hospitalizations related to obesity (and taxpayers footed half this bill through the government's Medicare and Medicaid medical programs).[3] In a September 2005 editorial in *The American Journal of Medicine,* Joseph S. Alpert, MD, and Pamela J. Powers, MPH, lamented: "We have a major health crisis in the United States: Americans are far too fat."[4] In September 2003, the Gallup Organization conducted a telephone interview of 301 female obstetrician-gynecologists. Out of ten, eight (78 percent, to be exact) said that obesity is of great concern to them as a problem facing their patients.[5]

Some of the most successful weight-loss programs employ teams of physicians, psychologists, and dietitians to keep their programs up to date and offer clients the online services they expect (food product ordering, recipes, menu plans, exercise programs, one-on-one nutritional counseling, online support groups, discount coupons for products and in-person meetings, and success stories).

Warning: although most of the programs covered in this chapter include a variety of food groups and encourage physical activity, everyone should

Internet Resources on Weight Loss and Obesity
© 2007 by The Haworth Press, Inc. All rights reserved.
doi:10.1300/5398_08

have a physical examination and seek advice from a healthcare practitioner *before* beginning a weight-loss regimen. A few of the programs are controversial, even extreme, in that they advise the exclusion of one or more food groups or advise dieters to base their meals around one food group. Be especially cautious about adopting one of these programs.

Herbal and pharmaceutical weight-loss products are addressed in another chapter, but their use warrants special caution due to possible side effects, interactions with prescription medications, and (in the case of herbal preparations) few rigorous scientific research studies on their effectiveness.

GENERAL OBESITY AND WEIGHT-LOSS GATEWAYS, GOVERNMENT SITES, AND ORGANIZATIONS

American Obesity Association
<www.obesity.org>

American Obesity Association's (AOA's) Web site is a portal to consumer information, legislation, research, statistics, tax and insurance issues, and discrimination related to obesity. Print fact sheets and read personal stories of patients. Law firms, pharmaceutical companies, food companies, health professional societies, and diet program companies support the site. AOA lists medical conditions linked to obesity, including nine types of cancer (breast, colorectal, endometrium, esophagus, gallbladder, liver, pancreas, renal cell, and top of the stomach).

American Society of Bariatric Physicians
<www.asbp.org>

American Society of Bariatric Physicians (ASBP) was founded in 1950. Membership is limited to licensed physicians. The mission of ASBP is "to advance and support the physician's role in treating overweight patients." The society offers continuing medical education courses for its members. The site allows consumers to locate member physicians in their home cities. A FAQ section provides information on obesity and its treatment, descriptions of weight-loss medications, statistics on prevalence of obesity in the United States, and documentation of the medical complications of obesity.

BestDietForMe.com
<www.bestdietforme.com>

For readers having difficulty choosing from among so many weight-loss diets, this site will help narrow the options. Marketdata Enterprises, a research publishing and consulting firm, created BestDietForMe, a diet analysis tool. The service is free (advertisers subsidize the site), and the questionnaire only requires a few minutes to complete. The questions cover age, weight, gender, height, budget preference, special psychological eating issues, and other personal habits, attitudes, and requirements as they affect selection of a diet program. BestDietForMe.com covers dozens of popular US-based diet programs. A personalized diet analysis appears on the Web site as soon as the reader completes the survey. Depending on responses to the questions, BestDietForMe offers diet programs for consideration. A dietitian and a clinical psychologist developed the diet survey.

Even if you have already selected a weight-loss program or are currently on a diet that is satisfactory, the information on weight loss, diet fraud, dieting tips, eating disorders, and weight-loss associations available at this site will be useful. Some of the categories are "consumer protection," "hot new topics," "factors affecting weight loss," "dieting for events," and several medical topics related to dieting.

The Diet Channel
<www.dietchannel.com>

This site won *Forbes Magazine's* Best of the Web award in the area of diet and nutrition. This is the place to go for brief descriptions of over sixty diets. It covers the spectrum from traditional (Weight Watchers) to faddish (Chocolate Diet) to bizarre (Russian Air Force Diet). An interesting feature is the section offering links to articles on diet and health topics such as phytochemicals, diet and cancer, antioxidants, and fad diets.

Healthy Weight Network
<www.healthyweight.net>

Be sure to read "How to Identify Weight Loss Fraud." The guidelines are extensive. There is also a section on how to report fraud. The site lists several questionable weight-loss pills, gadgets, and theories (although some practitioners and followers might take exception to the dismissal of hypnosis).

MayoClinic.com: Popular Diets:
The Good, the Fad, and the Iffy
<www.mayoclinic.com>

Select "Food & Nutrition Center" to read an article offering sound tips on evaluating several types of popular diets: low fat, low carbohydrate, glycemic index diets, meal replacements, and commercial diet programs.

National Heart, Lung, and Blood Institute:
Obesity Education Initiative
<www.nhlbi.nih.gov/about/oei/index.htm>

The Obesity Education Initiative makes key recommendations to the public on obesity, stressing the medical benefits of weight loss, the correct way to diet, and the importance of physical exercise (Figure 8.1).

FIGURE 8.1. Obesity Education Initiative
<www.nhlbi.nih.gov/about/oei/index.htm>

NAASO, the Obesity Society
<www.naaso.org>

NAASO is the leading U.S. organization for the scientific study of obesity. The society maintains statistics on the problem and promotes advocacy, research, and education. NAASO keeps physicians, scientists, and the public aware of new developments in the treatment and prevention of obesity. Read the fact sheets on obesity, childhood obesity, obesity and cancer, and obesity and diabetes.

Safediets.org
<www.safediets.org>

This is the weight-loss diet site of Physicians Committee for Responsible Medicine (www.pcrm.org), a nonprofit organization supported by physicians, nutritionists, and laymen who advocate preventive medicine and research, ethical treatment of human research subjects, alternatives to animal experimentation, and a low-fat, vegetarian diet. The site includes a three-week weight-loss plan consisting of meals built around four food groups: legumes, whole grains, fruits, and vegetables. Print out and duplicate the diet fact sheet, as it makes a very convenient daily diary for checking off required servings of the food groups as they are consumed each day. There is also a section rating the top diet books, but at the time we go to press, this does not mention the popular *South Beach Diet* book.

Shape Up America

See **Shape Up America! Pediatric BMI Assessment Tool** in Chapter 5.

Sisters Together: Move More, Eat Better
<www.niddk.nih.gov/health/nutrit/sisters/sisters.htm>

Sisters Together, a program of the National Institutes of Health's Weight-Control Information Network, features several online publications on its Web site: *Active at Any Size, Nutrition and Your Health, Weight Loss for Life, Sisters Together Program Guide,* and *Improving Your Health: Tips for African American Men and Women,* and other materials from the program.

US Food and Drug Administration:
Overweight and Obesity
<www.fda.gov>

The Food and Drug Administration (FDA) provides reliable material on all aspects of obesity. The site includes dozens of links to resources produced by the Centers for Disease Control, the FDA, National Institutes of Health, Federal Trade Commission, and Health and Human Services. Under "How to Lose and Manage Weight," the FDA advises readers to "watch your calories, be active." Be sure to read the FDA's plan of action to combat the problem of overweight, which includes improved food labeling to highlight calorie count, increased enforcement of companies that use inaccurate labeling, increased consumer education, improved guidelines for developing obesity drugs, encouraging restaurants to provide nutritional information, and working collaboratively with researchers, academia, government agencies, and nonprofit organizations to conduct obesity research.

The V8 Diet Plan
<www.v8juice.com>

The V8 Diet Plan is a balanced diet built around V8 vegetable juice and V8 Splash, combined with sensible meals. There is a diet plan and a chart with suggested substitutions (e.g., drink a Splash instead of a Frappuccino and save 180 calories).

Weight-Control Information Network
<win.niddk.nih.gov/index.htm>

Weight-Control Information Network (WIN), part of the National Institutes of Health, produces, collects, and disseminates information targeted at consumers and health professionals on different aspects of obesity, weight control, and nutrition. In addition to brochures and fact sheets (available online), WIN produces media programs, and has established Clinical Nutrition Research Centers (CNRCs) and Clinical Nutrition Research Units (CNRUs) to advance obesity and nutrition research. Be sure to read *Choosing a Safe and Effective Weight-Loss Program. Weight-Loss and Nutrition Myths: How Much Do You Really Know?;* intersperses myths about diet, meals, physical activity, and food with practical tips (Figure 8.2).

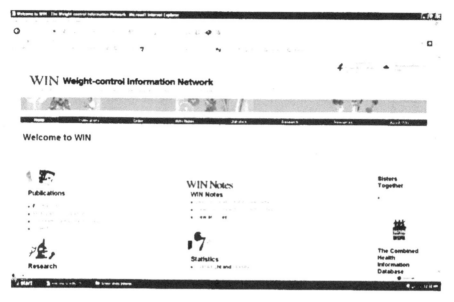

FIGURE 8.2. Weight-Control Information Network
<win.niddk.nih.gov/index.htm>

COMMERCIAL WEIGHT-LOSS PROGRAMS

Cyberdiet @ DietWatch
<cyberdiet.com>

DietWatch, which acquired Cyberdiet in 2000, is the largest Internet diet site. Paid subscribers to Cyberdiet receive a customized diet plan, meal plans, access to the online chat group, progress reports, shopping lists, fitness plans, and information on diet and nutrition. The site has won awards from *Forbes, PC World,* and About.com. Timi Gustafson, a registered dietitian, and Cynthia Fink developed CyberDiet in 1995.

Diet Center
<www.dietcenter.com>

The Diet Center is a franchise featuring a weight-loss program that emphasizes close, one-to-one supervision, dietary supplements, and meal replacement bars and drinks. The company has been in existence for over thirty years, so the program must succeed with some dieters.

Dr.Phil.com
<www.drphil.com>

Weight loss is just one facet of psychologist Dr. Phil McGraw's Web site. Select "Weight" from the listing of topics. This is a commercial site, with ads for his books and television show evident throughout. Weight-control advice focuses around seven steps developed to modify behaviors and emotions and to understand the psychological aspects of overeating. Unique features are the dieting readiness profile, the "Personal Environmental Audit," and the "Behavioral Audit."

eDiets
<www.ediets.com>

eDiets is a commercial portal to two dozen well-known diet plans. Its motto is "your diet, your way." Sign in, select a diet from the list, indicate a goal weight and a target date, fill in some basic facts (name, address, e-mail address, height, weight, gender, and age), bypass or choose to receive various electronic newsletters and product offers, and receive a free diet analysis. The site visitor immediately receives (online) a graph showing current weight and the target date and weight, as well as a daily caloric guideline. Another graph shows BMIs and explains health consequences if the BMI level is in the overweight or obese category. At this point, a registered dietitian presents the option of purchasing an online diet plan, cancelable at any time by calling an 800 number. Note that there is a cancellation fee if cancelled during the first three months. There are optional add-ons, such as professional support and access to Bob Greene's fitness training plan, which can significantly increase the weekly fee. Carefully read the fine print explaining how the fees are charged to a credit card. The basic membership fee includes weight-loss tracking, shopping lists, dining out and fast food advice, recipes, and menus. If a dieter needs a lot of personal attention, but cannot meet in person with a diet counselor, eDiets may be the ideal option.

Jenny Craig
<www.JennyCraig.com>

Jenny Craig diet program is known for its commercials featuring celebrity endorsements. The twenty-year-old franchised diet program emphasizes personal counseling and prepackaged, portion-controlled meals ("Jenny's Cuisine"). The Web site offers free recipes, a weight tracker, and advice, but consumers must sign up to receive the information. Jenny Craig also participates in eDiet's premium (fee-based) online diet program. There are Jenny Craig diet centers ("In-Centre Programs") in the United States, Canada, Puerto Rico, Australia, and New Zealand. Jenny Direct delivers prepackaged meals to the home and offers weekly telephone counseling and an online support group.

LA Weight Loss Centers
<www.laweightloss.com>

The LA Weight Loss Company began in 1989. Today, there are over 700 franchises worldwide. The Web site does not provide details of the weight-loss program, but LA Weight Loss sells optional prepackaged foods, there are no group meetings, and the diet does not require calorie counting. Members do not need to make appointments to come in for one-on-one counseling. Users are encouraged to visit a center or call an 800 number to enroll. There are special features on the site that are reserved for members. Free features include recipes, a BMI calculator, and an article from the American Heart Association offering ten tips for adding exercise to your daily routine.

Physicians Weight Loss Centers
<www.pwlc.com>

Physicians Weight Loss Centers, a franchised program of medically supervised diets, started in 1979. Today there are five weight-loss systems and "in-center" members and "online" members. The five weight-loss systems include programs for people needing to lose ten pounds or less, a more aggressive diet for those with larger weight-loss goals, a diet consisting entirely of Medifast meal replacement products, a diet that is built around meals dieters prepare themselves, and a diet supplemented with weight-loss medications such as Meridia and Xenical. There are centers in approximately twelve states. Blood work is required, and licensed physicians perform EKGs. There is one-on-one support. Staff carefully monitors changes in body measurements because sometimes the most dramatic changes are reflected in inches, not pounds.

Richard Simmons
<www.richardsimmons.com>

Richard Simmons has had a successful television show, a gym in Beverly Hills, California, has produced exercise tapes, and published diet aids (Deal-A-Meal and Food Mover). The Web site invites readers to join the "Clubhouse" in order to participate in online chats and other support options. The online store sells tapes and other Simmons products. The diet is based on motivation, exercise, and the American Dietetic Association/ American Diabetes Association food-exchange system.

WebMD Weight-Loss Clinic
<diet.webmd.com>

A nutritionist customizes an eating plan based on personal weight-loss requirements and current eating habits and food preferences. There is a weekly fee, charged quarterly to a credit card. Members have access to an online community and an interactive journal for tracking food consumption and weight-loss progress.

WeightWatchers.com
<www.weightwatchers.com>

The Weight Watchers Web site includes free information about the diet programs and traditional meetings; a message board; community recipe exchange; a few free Weight Watchers recipes; Weight Watchers Online, a subscription-only resource; and Weight Watchers eTools, a subscription-only Internet companion to traditional meetings.

There are traditional Weight Watchers meetings all over the world, and the Web site makes it easy to locate local meetings. The diet program regularly adds innovative enhancements (new ways to track flexible points and a "no counting" plan option), and the techniques used by meeting instructors have changed to encourage more member participation through meal planning exercises and questions presented for group discussion.

FASTING

Calorie Restriction Society
<www.calorierestriction.org>

The Calorie Restriction (CR) Society is a nonprofit organization advancing the practice of caloric restriction as a means to enhance health and longevity. The Web site hosts e-mail discussion groups, encourages scientific research on CR, and educates the public about the diet. CR practitioners restrict caloric intake while consuming an adequate supply of vitamins, minerals, and other nutrients. CR followers avoid simple sugars and flour while consuming nutrient-dense foods, primarily vegetables. Fats and proteins are chosen carefully and in moderate amounts. Roy L. Walford, MD, who researched the concept while at the University of California at Los Angeles, promoted Caloric Restriction with Adequate Nutrition (CRAN). In the mid-1980s and mid-1990s, he published *The 120 Year Diet* and *The Anti-Aging Plan*. In 2005, Walford's daughter, Lisa Walford, and Brian M. Delaney, president of the Caloric Restriction Society, published *The Longevity Diet*. The motto is "fewer calories, more life".

Medifast
<www.medifast1.com>

Medifast began in 1972 as the Nutrition Institute of Maryland. At one time, the Medifast protein-sparing fast was only available under close medical supervision in hospitals or doctors' offices. Now, most clients purchase the Medifast products directly from the company. The very-low-calorie diet consists of soy-based shakes and other meal replacement foods that are consumed along with one "lean and green" meat and vegetable meal daily. There are separate programs for men and women and a diabetic weight-loss program. The products meet kosher and lactose-free requirements.

Optifast Medical Weight-Loss Program
<www.optifast.com>

Optifast, a medically supervised program offered by Novartis Medical Nutrition, is designed for the clinically obese patient. There is a special regimen for bariatric surgery (weight-loss surgery) patients, both before and after surgery. Optifast formula was first developed in 1974. The program is offered only through hospitals and physicians' offices. Consumers can get local referrals on the Web site. A convenient feature of the site is the "toolbox," which pulls together links to nutrition, wellness, goal setting, fitness, and journaling (keeping a diary to record feelings and motivations).

HYPNOSIS

American Society of Clinical Hypnosis
<www.asch.net>

American Society of Clinical Hypnosis (ASCH) is one of two national organizations (the other is the Society for Clinical and Experimental Hypnosis) for licensed health professionals using hypnosis as a therapeutic modality. Hypnosis is an altered state of consciousness, a deep inner absorption with heightened concentration. Some individuals are more hypnotizable than others are. Use this site to locate hypnotherapists in your city. Note the sections "Selecting a Qualified Hypnotherapist," "Myths about Hypnosis," and "When Will Hypnosis Be Beneficial?" There is a distinction between lay hypnotists and licensed healthcare professionals (physicians, psychologists, dentists, nurses, and social workers) who use hypnosis as one of many treatment methods. ASCH recommends that consumers seek hypnotherapy from licensed healthcare professionals (Figure 8.3).

Society for Clinical and Experimental Hypnosis
<www.sceh.us>

Society for Clinical and Experimental Hypnosis (SCEH) is an international organization founded in 1949. Its members are health care professionals researching and practicing the medical use of hypnosis. SCEH publishes *International Journal of Clinical and Experimental Hypnosis,* a respected scholarly journal available in many medical school libraries. The organization offers workshops and symposia for healthcare practitioners wishing to further their hypnotherapy skills. Consumers can contact the society for a list of hypnosis professionals.

FIGURE 8.3. American Society of Clinical Hypnosis
<www.asch.net>
Reprinted with permission of American Society of Clinical Hypnosis.

LOW-CARBOHYDRATE DIETS

Atkins Nutritionals
<atkins.com>

The Atkins low-carbohydrate diet has been modified over the years, so the Web site plays an important role in keeping its followers informed of the changes. The Web site is one of the best of its type because many of the features are free and the site is kept current. It bridges the gap between the content contained in the Atkins diet books and the reality of going on a diet and facing questions or problems not addressed in the books. Atkins University, a section on the Web site, contains a support group and free, online "courses" on a variety of nutrition topics. The success stories of Atkins diet followers are varied and inspirational. The company has refocused since the death of Dr. Robert Atkins, with new books (dealing with diabetes while following the Atkins diet) and expansion of its line of food and nutritional products. Dieters can have meals delivered to their homes in some parts of the country ("Atkins At Home"). Portions of the site are commercial, with ads for books, foods, and links to the largely fee-based eDiets Web site, but the basic sections contain helpful information for followers of the program. Atkins Nutritionals Web site features news stories, events, and recipes. Sign up to receive a free e-mail newsletter. There is an online support group, and readers can contact Atkins Nutritionals information representatives by e-mail or by telephoning the Atkins 800 number. The Atkins Physicians Council has developed the Atkins Lifestyle Food Guide Pyramid, which differs from the US Department of Agriculture (USDA) MyPyramid, with protein sources (fish, beef, poultry, pork, and soy products) at the base, and whole-grain carbohydrates (barley, oats, and brown rice) at the apex. Overall, this is a well-designed, colorful site with features to meet every need. The company filed for Chapter 11 bankruptcy in mid-2005, but the reorganization of its debts does not seem to affect consumers. The company plans to concentrate on its successful line of foods, including nutrition bars, cereals, sauces, and chips. (*See also* **Carbohydrate Addict's Official Web Site, South Beach Diet Online,** and **The Zone Diet.**)

Carbohydrate Addict's Official Web Site
<www.carbohydrateaddicts.com>

This is the Web site of Drs. Richard and Rachael Heller, whose books, including *The Carbohydrate Addict's Cookbook,* have long been popular. The Hellers define carbohydrate addiction as the overwhelming craving for carbohydrate-rich foods, caused by the body's excessive release of insulin (hyperinsulinemia). The carbohydrate addict gains weight easily and has frequent swings in blood sugar levels. The most useful part of this site is the FAQ section, but the Web site appears to be designed primarily to sell the books.

Dr. Sears' Zone Diet *See* The Zone Diet.

South Beach Diet Online
<www.southbeachdiet.com>

Miami cardiologist Arthur Agatston, MD, developed the diet for his heart patients. He has since published best-selling books on the diet. The South

Beach Diet is built around nonstarchy vegetables, small amounts of fruit, low-fat protein, olive and canola oils, and carbohydrates with low-glycemic-index ratings, such as whole- grain breads and pasta. There are three phases to the diet, with carbohydrates and fruit added gradually during the second phase. Members, who are billed quarterly, have access to Q&A, chats, a weight tracker, meal plans, an online journal, message boards, Beach Buddies (dieters matched to other members for support), advice from online dietitians, food guides, and over 900 recipes. (*See also* **Atkins Nutritionals, Carbohydrate Addict's Official Web Site,** and **The Zone Diet.**)

The Zone Diet
<www.zonediet.com>

The Zone diet, developed by former Massachusetts Institute of Technology biotechnology researcher Dr. Barry Sears, consists of eating moderate amounts of carbohydrates, protein, and fats at every meal and snack. The Zone proportions for the nutrients are 40-30-30, with the goal of keeping insulin levels even. Dr. Sears' book *The Zone* was published in 1995. Prepackaged meals and supplements are sold at the Web site, but they are not a required part of the diet. Members receive access to support, a diet journal, an online community, dieting tips, and customized shopping lists and recipes. (*See also* **Atkins Nutritionals, Carbohydrate Addict's Official Web Site,** and **The South Beach Diet.**)

MEAL REPLACEMENT PRODUCTS
AND MEAL PURCHASE PROGRAMS

Herbalife
<www.herbalife.com>

Herbalife is a twenty-five-year-old company that sells weight-loss and nutrition/wellness supplements and meal replacement products through a network of independent distributors or ShapeWorks coaches. To purchase any of the products described on the Web site, fill out a form (you are required to list your telephone number) for referral to a local distributor. The prices are not posted on the site.

NutriSystem Nourish
<www.nutrisystem.com>

NutriSystem has been around for over thirty years, but the program is now Web based. There are no longer any NutriSystem diet centers. Dieters purchase meal replacement products (packages for men, women, type 2 diabetics, and vegetarians) through the Web site and receive counseling online or via the telephone. Membership is free and includes access to on-line chat groups, a newsletter, online classes, and unlimited counseling. Meal replacement products are supplemented with vitamins and fresh fruit and vegetables. The NutriSystem meals include "good carbs," carbohydrates with a low glycemic index. Low-glycemic-index carbohydrates are metabolized slowly and do not cause spikes of insulin level (Figure 8.4).

FIGURE 8.4. NutriSystem Nourish
<www.nutrisystem.com>
Reprinted with permission of NutriSystem Nourish.

Slim-Fast.com
<www.slim-fast.com>

Slim-Fast offers four versions of the Optima Diet, combining meal replacement shakes and bars with sensible meals and snacks. Three of the plans are geared to the dieter's starting weight, while the fourth option is a low-carb diet. The products are carried in supermarkets and drug stores. The Web site offers plenty of dieting advice, online chats, meal planning guidelines, and recipes.

PSYCHOTHERAPY

See **Dr. Phil.com**

APA Online *Monitor*
<www.apa.org/monitor/jan04/bringing.html>

An article in the American Psychological Association's publication, *Monitor* (electronic version), gives a thorough overview of methods that psychologists employ to help patients lose weight. "Bringing More Effective Tools to the Weight-Loss Table" lists self-monitoring, meditation, and accentuated cognitive behavior therapy (CBT). With accentuated CBT, antidepressant medications, liquid diets, bariatric surgery, and/or weight-loss drugs are also employed, because CBT alone has a high failure rate when used for weight loss.

SPIRITUALITY- AND RELIGION-BASED WEIGHT-LOSS PROGRAMS

First Place: The Bible's Way to Weight Loss
<www.firstplace.org>

First Place began in 1981 as a ministry of First Baptist Church of Houston. There are church-based groups worldwide. Participants are encouraged to exercise at least three times a week, follow a sensible diet based on the six food groups, engage in Bible study (read two Bible chapters daily, pray daily, and memorize one Bible verse weekly), attend one weekly First Place meeting, and keep a commitment food diary. Members and group leaders must purchase kits containing pamphlets, the Live-It food plan, and DVDs (leader's kit) and CDs (member's kit). The leader's kit costs approximately $180 and the member's kit is approximately $80. The Web site sells the kits and books, Bible study materials, and accessories with the First Place logo.

Spirituality & Health Magazine's The Solution
<www.spiritualityhealth.com>

The Solution is a free online weight-loss group developed and moderated by Laurel Mellin, MA, RD, associate clinical professor of family and community medicine at the University of California San Francisco School of Medicine. Participants in the six-week "e-course" receive e-mails from Mellin encouraging them to exercise more and to reflect about issues related to eating, such as emotional hunger versus physical hunger. Group participants do not interact with each other. The e-course is an introduction to techniques explored in greater depth in Mellin's book *The Solution: Never Diet Again* and in articles in *Spirituality & Health*, a print magazine. *Spirituality & Health* does not reflect any particular religious doctrine. The editors draw from all major religions and combine spiritual practice with insights from science and medicine. The Web site features several other e-courses in addition to The Solution, and readers can take several online self-tests. Two, The Integrative Eating Program: Your Personal Eating Profile and Weight Loss complement The Solution.

The Weigh Down Workshop
<www.wdworkshop.com>

Gwen Shamblin, an RD and university nutrition instructor with experience as a weight-loss consultant, founded the Christianity-based diet program in the 1980s. Members purchase syllabi for courses that are offered online or on DVDs. Members, who recruit five or more participants, receive a free copy of the syllabus.

MISCELLANEOUS

Sugar Busters! Diet
<www.sugarbusters.com>

The mantra of this program is "cut sugar to trim fat." Three physicians, Samuel S. Andrews, Morrison C. Bethea, and Luis A. Balart, and a successful follower of the diet, H. Leighton Steward, wrote the *Sugar Busters!* book. The key concept of the diet is that consumption of refined sugar and bad carbohydrates (processed grain products) causes the body to produce more insulin, store fat, and gain weight. On the Sugar Busters! program, dieters are encouraged to eat fruit, high-fiber vegetables, low-fat meats and fish. Exercise is an important component of the program. The Web site sells frozen meals, snacks, beverages, and sugar substitutes.

Trevose Behavior Modification Program
<www.tbmp.org>

Shape up or ship out! A tough-love attitude may be what a dieter needs to finally get serious about weight loss. Trevose Behavior Modification Program (TBMP) was started in Philadelphia in 1970. Insurance executive David S. Zelitch developed the program with the direction of world-famous obesity authority Dr. Albert J.Stunkard of the University of Pennsylvania's School of Medicine. In order to apply to join the program, consult the Web site for satellite meeting groups that have openings for new participants. Once there is an opening, send a letter stating what you have done in the past to lose weight and why you believe you are ready to lose weight now. Include a self-addressed, stamped envelope. Most members wait three to six months before they are admitted into the program. There are no fees, but participants are encouraged to make contributions to cover the cost of supplies. There is a newsletter. Participants are required to lose a predetermined amount of weight each month. They must attend every weekly meeting. The emphasis is on behavior modification, and the first six months on the program are devoted to studying participants' eating habits. Read the "Rules" section of the Web site to learn all the weight-loss and attendance requirements. The program only offers satellite meetings in the Delaware Valley (southeastern Pennsylvania and southern New Jersey). Successful participants are encouraged to give back to TBMP by becoming trained meeting leaders. Anyone who "flunks" out of the program is not permitted to reapply.

NOTES

1. Lissau, I., Overpeck, M.D., Ruan, W.J., Pernille, D., Holstein, B.E., Hediger, M.L., and the Health Behaviour in School-Aged Children Obesity Working Group. (2004). Body Mass Index and overweight in adolescents in European Countries, Israel, and the United States. *Archives of Pediatrics and Adolescent Medicine* 158:27-33.

2. Sturm, R. (2002). The effects of obesity, smoking, and drinking on medical problems and costs. *Health Affairs* 21:245-253.

3. Finkelstein, E.A., Fiebelkorn, I.C., and Wang, G. (2004). State-level estimates of annual medical expenditures attributable to obesity. *Obesity Research* 12:18-24.

4. Alpert, J.S. and Powers, P. J. (2005). Editorial: Obesity: a complex public health challenge. *American Journal of Medicine* 118:935.

5. American College of Obstetricians and Gynecologists. ACOG News Release: Female Ob-Gyns name obesity the greatest threat to women's health. 9 December 2003. Available: http://www.acog.org/from_home/publications/press_releases/nr12-09-03-1.cfm Accessed 5 March 2006.

Chapter 9

Weight-Loss (Bariatric) Surgery
Web Sites

Bariatric (from *baros*, the Greek word meaning "weight" and *iatrikos*, meaning "art of healing") surgery involves reducing the capacity of the stomach or the absorption of food by means of one of several procedures. Edward E. Mason, MD, and C. Ito, MD, of the University of Iowa,[1] first performed gastric bypass in the 1960s. In 2003, an estimated 100,000 people had bariatric surgery, and the number exceeded 200,000 in 2004. Of all American adults, 6 to 10 percent are considered to be morbidly obese, having a body mass index (BMI) greater than forty. Bariatric surgery is usually limited to people in this category, unless they also have major, life-threatening diseases (comorbidity factors), such as poorly controlled hypertension (high blood pressure), diabetes, severe heart or lung problems, or obstructive sleep apnea. Patients with these complicating medical conditions are usually eligible for bariatric surgery when their BMIs are in the range of 35-40.

A psychological evaluation is often part of the presurgical process. Patients are poor candidates for surgery if the screening reveals signs of major depression or other serious psychological issues. The prospective patient must understand the risks of bariatric surgery. In the United States, 1 of every 200 people who undergo this surgery dies within thirty days of the procedure.[2] Serious complications following surgery include lung blood clots (pulmonary emboli), wound infections, bleeding, and gastrointestinal leakage. The benefits of major abdominal surgery must exceed the risks. Patients must understand that their responsibilities do not end with the surgery: they must commit to a lifetime of carefully monitoring food intake

Internet Resources on Weight Loss and Obesity
© 2007 by The Haworth Press, Inc. All rights reserved.
doi:10.1300/5398_09

and maintaining general health and vitamin and mineral status (anemia and calcium, iron, or vitamin B_{12} deficiencies are common). Patients are cautioned to avoid foods high in fat or refined sugar, lest they develop "dumping syndrome" (a group of unpleasant symptoms including dizziness, nausea, diarrhea, and rapid heart beat). After bariatric surgery, the stomach is much smaller and may hold only 2 to 6 ounces of food at one time (a normal stomach holds 40 to 50 ounces). Patients are cautioned to avoid drinking fluids during meals and are advised to chew food until it is almost liquid. Dry foods, such as nuts, bread, and popcorn may stick and cause vomiting. High-fiber foods can cause cramping. There may be cosmetic issues to resolve after losing a great deal of weight: sagging skin on the face and body. Some patients will require plastic surgery to deal with hanging folds of skin on the arms, neck, and abdomen. Glance through any plastic surgery journal to see photographs of cosmetic surgery on arms, legs, chins, bellies, and even genitalia of patients who have lost dramatic amounts of weight following surgical procedures. Late-term complications, such as intestinal blockage due to rearrangement of the intestines during surgery, may occur. Whenever someone undergoes a major lifestyle change, issues may arise with family members and friends. Spouses may feel insecure or jealous if the patient becomes physically attractive to other people. The patient may have problems adjusting to his or her new appearance or may continue to view him or herself as obese (body image disturbances). The actual weight loss or change in appearance may not be what was expected. Mattison and Jensen[3] report, "Patients must realize that the definition of success for these procedures is the loss of 50 percent of their excess weight . . . it seldom results in the patient achieving his or her ideal weight." The weight will come back if the patient does not exercise or eats small amounts of fattening food all day long ("outeating").

In the United States, the most popular types of bariatric surgical procedures are laparoscopic gastric banding (Lap-Band adjustable gastric band), including vertical-banded gastroplexy (VBG), and gastric bypass (Rouxen-Y gastric bypass), representing the categories "restrictive" and "malabsorptive," respectively. With restrictive procedures, food is absorbed and digested normally, but the patient eats less—he or she quickly feels full because the stomach is smaller. With malabsorptive procedures, less food is absorbed due to the bypassing of lengths of the small bowel. Currently under evaluation is a safer, two-stage procedure where patients first undergo laparoscopic sleeve gastrectomy, in which a large portion of the stomach is removed. Months later, they undergo the Roux-en-Y procedure, which involves creation of a small stomach pouch and bypassing of a portion of the intestines.[4,5] Procedures cost an average of $25,000.

See the following Web sites for details on specific bariatric procedures.

American Society for Bariatric Surgery
<www.asbs.org>

American Society for Bariatric Surgery, ASBS, a professional organization of physicians specializing in weight-loss surgery, was established in 1983. Its mission is to educate and train physicians, keep the public and allied health professionals updated about this type of surgery, set guidelines for the ethical selection and treatment of patients, exchange knowledge and information among its members, conduct research, and promote quality assurance and evaluation of surgical outcomes. The Web site features a BMI calculator, lists of continuing medical education courses for physicians, "Rationale for the Surgical Treatment of Morbid Obesity," and a lengthy history of the evolution of bariatric surgery.

Bariatric Surgery Center at Highland Hospital
<www.stronghealth.com/services/surgical/bariatric>

Highland Hospital and Strong Memorial Hospital are affiliated with University of Rochester Medical Center. The section "Diet" under "Understanding Bariatric Surgery" is extensive, featuring several postoperative gastric bypass diets. Each diet includes detailed instructions and charts showing allowed foods, foods to avoid, and miscellaneous tips.

BariatricEating.com
<www.bariatriceating.com>

Susan Maria Leach, a woman who has successfully maintained her weight and good health several years after bariatric surgery, developed and maintains this Web site. She owns BariatricEating, a company that sells protein bars and shakes and vitamin supplements for bariatric surgery patients. Leach authored a cookbook *Before & After: Living & Eating Well*

After Weight Loss Surgery, which is sold on the Web site. BariatricEating. com includes a message board, recipes, before and after photographs of patients, and a special Bariatric Help Center, which includes areas for "pre-ops" and "post-ops" patients. "Ask Susan Maria," links to American Society for Bariatric Surgery centers of excellence, recipes, and information on specific weight-loss surgery procedures (Figure 9.1).

BariatricEdge
<www.bariatricedge.com>

Ethicon Endo-Surgery, Inc., a manufacturer of surgical knives, staplers, and other devices used during surgery, maintains BariatricEdge. Type in a zip code to see a list of bariatric surgeons in your city. These surgeons meet rigid criteria relating to experience, training credentials, and quality of ancillary staff and hospital facilities. The section on risks of surgery is extensive and easy to understand.

FIGURE 9.1. BariatricEating.com
<www.bariatriceating.com>
Reprinted with permission of BariatricEating.com.

Gastrointestinal Surgery for Severe Obesity
<www.niddk.nih.gov/publications/gastric.htm>

This is a section of the Weight-Control Information Network (WIN) of the National Institutes of Health's National Institute of Diabetes and Digestive and Kidney Diseases. WIN was established in 1994 to provide science-based information on obesity, weight control, and nutrition to consumers and health professionals. The site is one of the most extensive Web resources on bariatric surgery. There are illustrations, very clear and thorough descriptions of the advantages and drawbacks of all types of gastric bypass surgery, and a list of questions a patient should ask himself or herself before deciding to undergo bariatric surgery (Figure 9.2).

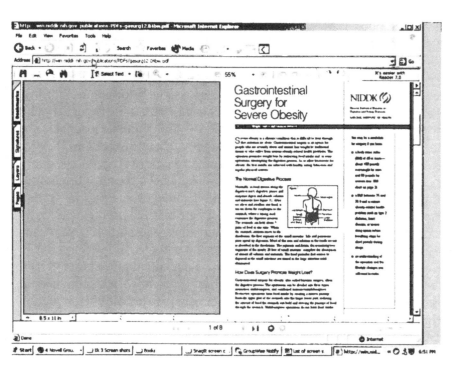

FIGURE 9.2. Gastrointestinal Surgery for Severe Obesity
<www.niddk.nih.gov/publications/gastric.htm>

MEDLINEplus.gov
<medlineplus.gov>

Select "Surgical Videos" to view videotapes of bariatric surgical procedures. Not for the squeamish, but readers will want to see what happens when the surgeon picks up his or her knife. Note that MEDLINEplus's section on obesity is an authoritative gateway to consumer-oriented information on all aspects of obesity and weight loss.

ObesityHelp.com
<www.obesityhelp.com/morbidobesity>

In 1998, Eric Klein, a former emergency medical technician, founded the Association for Morbid Obesity Research and the Web page Obesity-Help to provide peer support for the morbidly obese, including patients undergoing or contemplating bariatric surgery. This site has a commercial aspect, but ObesityHelp has so many unique informative sections that it should be a top selection for consumers seriously considering weight-loss surgery. Features include a clothing exchange, "Ask a Surgeon," "Find a Bariatric Surgeon," "Find a Plastic Surgeon," and "Find a Hospital." The "Bariatric Hospital Directory" is very useful: click on a state, and view information about local facilities that perform this type of surgery. The directory lists the number of procedures performed, date of the most recent procedure, accreditation of the hospital, and detailed evaluations by former patients. Readers of ObesityHelp can post personalized profiles and locate peers. A patient may elect to list his or her name, photograph, type of surgery, and date of scheduled surgery ("Recent and Upcoming Surgeries"). Peers are encouraged to provide support, especially the week before and after surgery. Reading the diarylike entries, especially the material about postoperative periods, should be of immense help to people contemplating weight-loss surgery. Physicians or nurses can never convey details of surgical experiences as vividly. There are several types of chat rooms ("forums"), including one on nutrition, with information contributed by dietitians. The description of several types of surgical procedures is excellent, with illustrations and the advantages and disadvantages displayed in chart format. Klein requests contributions and hospital/physician sponsorship to help defray costs. Readers can subscribe to a print magazine *(ObesityHelp)*. Over the years, there have been over 150,000 members.

ObesityHelp is the most valuable Web site in this category because of the accounts of patient experiences.

Obesity-Surgery.net
<www.obesity-surgery.net/obesity_surgery_faq.htm>

Obesity-Surgery.net is the Web site of a bariatric surgeon's clinic, but the FAQ section is clearly and honestly written. Be sure to read the information on weight loss and regain of lost weight after bariatric surgery.

Obesity Surgery Support Group
<groups.yahoo.com/group/ossg>

Obesity Surgery Support Group, OSSG, also called Planet OSSG, is a moderated discussion group for people who have had bariatric surgery or are contemplating having this surgery. To subscribe, read the instructions at this site.

Piedmont General & Bariatric Surgery.
Bariatric Surgery Diet Manual
<www.bariatricsolutions.com>

This is the Web site of a Greenville, South Carolina, medical practice specializing in the medical and surgical treatment of obesity. The diet manual serves patients before and after bariatric surgery, and covers vitamins, minerals, food intolerances, diet timetables, and nutritional problems.

Thinner Times Gastric Bypass
<thinnertimes.com>

Thinner Times is the Web site of a private surgical clinic, but the information about gastric surgery and the links to support groups are useful. Prospective patients will want to read the section on complications (select "Gastric Bypass" on the home page).

NOTES

1. Mason, E.E. and Ito, C. (1967). Gastric bypass in obesity. *Surgical Clinics of North America* 47:1345-51.

2. Windham, C. (2003). Gastric-bypass study shows risks of surgery. *Wall Street Journal* (Eastern edition), December 2, p. D6.

3. Mattison, R. and Jensen, M.D. (2004). Bariatric surgery: for the right patient, procedure can be effective. *Postgraduate Medicine* 115:49-58.

4. Gallagher, S. (2004). Taking the weight off with bariatric surgery. *Nursing* 34:58-63.

5. Steinbrook, R. (2004). Surgery for severe obesity. *New England Journal of Medicine* 350:1075-1079.

Chapter 10

Prescription Medications
for Weight Loss

Barbara's Obesity Meds and Research News
<www.obesity-news.com>

Barbara Hirsch, a medical journalist, is the developer and manager of this resource, which is aimed at both consumers and physicians. Portions of the site are only accessible to paid subscribers, but the "premium" sections are clearly color-coded. Read "Accuracy in Medical Reporting: A Consumer's Guide" for criteria on gaging the quality of medical articles. Hirsch has put together so much information on the topic of obesity, that serious obesity researchers and consumers might find the annual individual (U.S.) subscription rate of $104 a good value. Read the information on orlistat (Xenical), a drug that inhibits the absorption of fat (lipase inhibitor), and sibutramine (Meridia), a serotonin and norepinephrine reuptake inhibitor (these brain chemicals may have an effect on appetite and satiety). Hirsch also includes extensive information on new diet medications currently in the human testing stage before Food and Drug Administration (FDA) approval. In addition to safety and efficacy, she explains the factors that may affect continuation of clinical trials for a new drug.

National Association to Advance Fat Acceptance (NAAFA)
<www.naafa.org>

Read NAAFA's Policy on Weight-Loss Drugs, which includes a chronology of the development and use of weight-loss medications in the

Internet Resources on Weight Loss and Obesity
© 2007 by The Haworth Press, Inc. All rights reserved.
doi:10.1300/5398_10

United States. NAAFA contends that the FDA has failed to protect consumers from dangerous or ineffective weight-loss medications, and it needs to reform its drug approval process.

U.S. Food and Drug Administration:
Approved Obesity Treatments, Drugs,
FDA Talk Paper on Orlistat
<www.fda.gov/bbs/topics/ANSWERS/ANS00951.html>

The FDA releases consumer statements on approved drugs and surgical procedures. *FDA Approves Orlistat for Obesity* was released on April 26, 1999, but the information is still current. The FDA revises talk papers when new information becomes known. Orlistat is the generic name of Xenical, a lipase inhibitor. Lipase inhibitors prevent enzymes in the gastrointestinal tract from breaking down dietary fats into smaller molecules that the body can absorb. This results in calorie loss. The FDA statement advises orlistat users to take supplements of fat-soluble vitamins (A, B, E, and K), because this medication interferes with the absorption of these vitamins.

Weight-Control Information Network:
Prescription Medications for the Treatment of Obesity
<win.niddk.nih.gov/publications/prescription.htm>

This publication on the WIN Web site covers the two main types of obesity medications: lipase inhibitors and appetite suppressants. There is also information on "off-label" use of antidepressants, antiseizure medications, and diabetes drugs for the treatment of obesity. Read the warnings about use of drug combinations or amphetamines. The site includes information on medication side effects, and advises caution for obese patients with additional health problems.

Chapter 11

Weight-Loss Spas
and Residential Diet Programs

Canyon Ranch Health Resorts
<www.canyonranch.com>

Holistic medicine practitioner Jesse F. Williams, MD, influenced Mel Zuckerman, who founded Canyon Ranch in 1979. Williams' concept of health was not the mere absence of illness. Health is the condition that makes it possible for one to experience the greatest enjoyment of life. This is the philosophy of Canyon Ranch Health Resorts and Canyon Ranch Spa-Clubs. Today, there are Canyon Ranch Health Resorts in Tucson, Arizona, and Lenox, Massachusetts. The medical staff at both facilities comprises board-certified physicians, dietitians, physical therapists, exercise physiologists, behavioral health specialists, acupuncturists, and other health specialists. The Web site includes brief biographies for all of these clinicians.

Guests choose to focus on one of several health programs, including "Ultrametabolism: New Approaches to Weight Loss" and "Ultraprevention," which focuses on nutrition, fitness, and strategies to enhance longevity and energy. Guests attend hands-on cooking classes to learn healthy eating.

Miraval Life in Balance Resort
<www.miravalresort.com>

Miraval Resort is located in Tucson, Arizona. Center for Life in Balance programming includes a weight-loss program "Weight in Balance." Guests wishing to concentrate on nutrition or weight loss can attend cooking dem-

Internet Resources on Weight Loss and Obesity
© 2007 by The Haworth Press, Inc. All rights reserved.
doi:10.1300/5398_11

onstrations, dine with a nutritionist, and attend classes on food labeling, mindful eating, weight loss without dieting, food pantry reorganization, and mindful dining out.

Dr. Dean Ornish Program for Reversing Heart Disease
at Windber Medical Center
<www.windbercare.com/ornish.html>

This is one of several residential programs (one-day, three-day, and weeklong retreats) sanctioned by Dean Ornish, MD. It is recommended for patients contemplating coronary bypass surgery, who have experienced a cardiac event, who have coronary artery disease, or who have significant high blood pressure, high cholesterol, or obesity. Windber Medical Center is a Planetree-affiliated hospital in Somerset County, Pennsylvania. Planetree hospitals believe in the mind-body-spirit connection. Care is patient centered. The Ornish program features lifestyle changes that include a low-fat, whole-grain, vegetarian diet, moderate exercise, and stress reduction techniques. A registered dietitian manages the Windber program.

Pritikin Longevity Center and Spa
<www.pritikin.com>

Nathan Pritikin, Dr. David Lehr, and Dr. Robert Bauer founded the Pritikin Longevity Center in 1978. Today, Paul Bauer, son of Robert Bauer, directs the program in Aventura, Florida. The medically supervised program consists of exercise and a diet that is low fat, low salt, and low in refined carbohydrates. The diet is high in fiber, antioxidants, phytochemicals, vitamins, and minerals. The diet permits modest amounts of fish and low-fat dairy products. The program is successful in aiding weight loss, relieving the pain of arthritis, and controlling high cholesterol, atherosclerosis, and diabetes.

Rice Diet Program
<www.ricedietprogram.com>

The Rice Diet Program was developed in 1939 at Duke University in Durham, North Carolina. The residential "lifestyle" program specializes in treating patients with serious chronic illnesses such as hypertension, diabetes, kidney disease, heart disease, and obesity. Patients attend classes every day and are closely monitored by medical personnel. The Web site offers detailed information about the program, including fees and descriptions of the phases of the low-fat, low-salt diet. The diet emphasizes fruit and grains, gradually adding vegetables and fish. Most patients stay for four to eight weeks.

SpaFinder
<www.spafinders.com>

SpaFinder is a spa travel and marketing company that began in 1986. It publishes a book and a magazine in addition to the Web site. Select "Spas by Category" or "Medical Spa Program." SpaFinder covers international destination spas and day spas. Search by price, geographic location, and many other categories.

Chapter 12

Recipe Web Sites

For suitable recipes, also see the recipe sections in many of the general nutrition and diet program Web sites.

All Recipes
<www.allrecipes.com>

This site provides "Real recipes for real people." Search "Recipes Collection" for low-carb, low-fat, and vegetarian recipes. The section. "Healthy Solutions" has recipes for management of high blood pressure, high cholesterol, and general wellness. A year of customized meal planners ("Nutri-Planners") is available for purchase: Diabetes Management Program, Healthy Heart Program, or General Wellness Program.

Cook's Thesaurus
<www.foodsubs.com>

Cook's Thesaurus provides substitutions for recipe ingredients (ethnic substitutions, low fat, low cost, and low calorie).

Cooking Light
<www.cookinglight.com>

Cooking Light is the Web site of the magazine of the same name. It includes recipes plus Cooking 101, with a pantry ingredient list, food storage guide, and annual recipe index.

Internet Resources on Weight Loss and Obesity
© 2007 by The Haworth Press, Inc. All rights reserved.
doi:10.1300/5398_12

Flora's Recipe Hideout
<www.floras-hideout.com/recipes.index.html>

There are a dozen categories, with emphasis on desserts. The low-carb and low-fat categories are excellent and varied.

Meals for You
<www.mealsforyou.com>

Meals for You allows the reader to find recipes meeting common nutritional requirements: low fat, low calorie, low sodium, diet points (*a la* Weight Watchers), low carb, etc.

Recipe Archive Index
<www.cs.cmu.edu/~mjw/recipes>

Carnegie Mellon University's School of Computer Science hosts this site. The resource provides links to hundreds of recipes.

Recipe Link
<www.recipelink.com>

Recipe Link includes recipes from newspaper columns, recipes using brand name ingredients, kosher recipes, romantic meal recipes, brunch recipes, and recipes catagorized by the particular ingredients they use.

SOAR: Searchable Online Archive of Recipes
<www.recipesource.com>

Recipe Source hosts SOAR, a very extensive grouping of recipes. It is particularly strong in ethnic cuisine.

Glossary

activities calculator: Enter your current weight and the number of calories you wish to burn. This calculator suggests physical activities and how long it will take to burn off those calories.

anorexia nervosa: Endless dieting to the point of starvation. The patient intensely fears gaining weight and has a distorted perception of his or her body image.

bariatric surgery: Weight-loss surgery.

basal metabolic rate (BMR): Amount of calories the body burns when at rest, but awake, over the course of one day. *See also* CALORIE/ENERGY NEEDS.

binge eating (Compulsive Overeating): Consumption of large amounts of food frequently and repeatedly. Person feels out of control, may eat rapidly and secretly, and feels ashamed. Unlike bulimics, binge eaters rarely vomit or overexercise.

body mass index (BMI): A formula for standardizing the extent of obesity, based on body measurements. The formula: take your weight in pounds divided by your height in inches squared. Multiply this number by 703. A BMI between 19 and 24 is considered normal weight.

bulimia nervosa: From the Greek *boulimia,* meaning ravenous hunger. It is a syndrome of binge eating at least twice a week, followed by self-induced vomiting, fasting, purging, use of enemas, or compulsive exercising to avoid weight gain.

calorie/energy needs: Amount of calories a person needs to eat each day to maintain weight and fuel physical activity.

Internet Resources on Weight Loss and Obesity
© 2007 by The Haworth Press, Inc. All rights reserved.
doi:10.1300/5398_13

calories burned: It is the number of calories burned in an activity. Select type and duration of activity, and the calculator figures how many calories have been burned.

clinically severe obesity: Newer term for morbid obesity.

dumping syndrome: Group of unpleasant symptoms that may occur after weight-loss surgery because of the altered ability to digest sugars and fats and rapid movement of food from stomach to small intestines.

duodenum: First portion of the small intestine.

glycemic index (GI): A measurement of the impact a food has on blood sugar and insulin levels. Researchers at the University of Sydney developed this index. The GI is the rate of rise of blood sugar for a given food when compared with glucose, which has a GI of 100. Low-GI foods are thought to promote weight loss by providing a feeling of fullness (satiety) after a meal. Low-GI foods are thought to promote fat burning. Basmati rice has a low GI, while a baked potato has a high GI. Al dente spaghetti has a lower GI than spaghetti that has been boiled for 10 min. "Good carbs" have lower GIs than "bad carbs."

jejunum: Second portion of the small intestine.

ketosis: Process in which ketones are formed in the body during fasting.

morbid obesity: Condition where the BMI is 40 or greater. *See also* CLINICALLY SEVERE OBESITY.

obesity: Condition where the BMI is 30-39.9, indicating an abnormally high proportion of body fat.

overweight: Condition where the BMI is 25-29.9.

Roux-en-Y gastric bypass: Bariatric surgical procedure that combines restriction and malabsorption. Stapling or banding the stomach creates a small stomach pouch. A Y-shaped portion of the small intestine is attached to the pouch, bypassing the duodenum and a portion of the jejunum.

vertical banded gastroplexy (VBG): Bariatric surgical procedure that restricts the size of the stomach. A band and stapling create a small stomach pouch.

waist-to-hip ratio: Calculation used to determine whether body is pear or apple shaped.

Bibliography

American Heart Association. (2005). *The No-Fad Diet: A Personal Plan for Healthy Weight Loss.* New York: Clarkson Potter Publishers.

Blackwood, H. S. (2005). Help your patient downsize with bariatric surgery. *Med/Surg Insider.* Fall: 4-9.

Duyff, R. L. (2002). *American Dietetic Association Complete Food and Nutrition Guide* (Second Edition). Hoboken, NJ: John Wiley & Sons.

Hark, L. & Deen, D. (2005). *Nutrition for Life.* New York: Dorling Kindersley.

Katz, D. L. & Gonzalez, M. H. (2002). *The Way to Eat.* Napierville, IL: Sourcebooks.

Smolin, L. A. & Grosvenor, M. B. (2005). *Basic Nutrition.* Philadelphia: Chelsea House.

Internet Resources on Weight Loss and Obesity
© 2007 by The Haworth Press, Inc. All rights reserved.
doi:10.1300/5398_14

Index

Numbers followed by the letter "f" indicate figures.

T - #0605 - 101024 - C0 - 216/138/6 - PB - 9780789026507 - Gloss Lamination